The Montana Mathematics Enthusiast

Monograph 4

Creativity, Giftedness, and
Talent Development in Mathematics

The Montana Mathematics Enthusiast

Monograph 4

Creativity, Giftedness, and Talent Development in Mathematics

Edited by

Bharath Sriraman
The University of Montana

INFORMATION AGE PUBLISHING, INC.
Charlotte, NC • www.infoagepub.com

Library of Congress Cataloging-in-Publication Data

Creativity, giftedness, and talent development in mathematics / edited by Bharath Sriraman.
  p. cm. – (Montana mathematics enthusiast monograph series in mathematics education)
 Includes bibliographical references.
 ISBN 978-1-59311-977-5 (pbk.) – ISBN 978-1-59311-978-2 (hardcover)
1. Mathematics–Study and teaching (Elementary) 2. Mathematics–Study and teaching (Middle school) 3. Mathematics–Study and teaching (Secondary) 4. Gifted children–Education. 5. Creative thinking in children. I. Sriraman, Bharath.
 QA11.2.C74 2008
 510.71–dc22

                                        2008027251

Printed in the United States of America

To Robert Wheeler and Paula Olszewski-Kubilius
for nurturing and encouraging my interest
in creativity and giftedness

CONTENTS

Preface: Creativity, Giftedness, and Talent Development
in Mathematics ix
Bharath Sriraman

1   The Characteristics of Mathematical Creativity 1
Bharath Sriraman

2   Mathematical Giftedness, Problem Solving and the Ability
to Formulate Generalizations: The Problem-Solving
Experiences of Four Gifted Students 33
Bharath Sriraman

3   Gifted Ninth Graders' Notions of Proof: Investigating Parallels
in Approaches of Mathematically Gifted Students and
Professional Mathematicians 61
Bharath Sriraman

4   Are Mathematical Giftedness and Mathematical Creativity
Synonyms?: A Theoretical Analysis of Constructs 85
Bharath Sriraman

5   Does Mathematics Gifted Education Need a Working
Philosophy of Creativity? 113
Viktor Freiman and Bharath Sriraman

6   Enabling More Students to Achieve Mathematical Success:
A Case Study of Sarah 133
Sylvia Bulgar

7   Problems to Discover and to Boost Mathematical Talent in
Early Grades: A Challenging Situations Approach 155
Viktor Freiman

8 Mathematical Problem Solving Processes of Thai Gifted
Students 185
Supattra Pattivisan and Margaret L. Niess

9 Knowledge as a Manifestation of Talent: Creating
Opportunities for the Gifted..209
Alexander Karp

10 An Ode to Imre Lakatos: Or Quasi-Thought Experiments
to Bridge the Ideal and Actual Mathematics Classrooms 225
Bharath Sriraman

11 The Mathematically Gifted Korean Elementary Students'
Revisiting of Euler's Polyhedron Theorem.................................... 251
Jaehoon Yim, Sanghun Song, and Jiwon Kim

12 Mathematically Promising Students from the Space Age
to the Information Age... 271
Linda Jensen Sheffield

13 Revisiting the Needs of the Gifted Mathematics Students:
Are Students Surviving or Thriving? 277
Alan Zollman

14 Playing with Powers 287
Bharath Sriraman and Pawel Strzelecki

PREFACE

CREATIVITY, GIFTEDNESS, AND TALENT DEVELOPMENT IN MATHEMATICS

Bharath Sriraman
The University of Montana

Our innovative spirit and creativity lies beneath the comforts and security of today's technologically evolved society. Scientists, inventors, investors, artists and leaders play a vital role in the advancement and transmission of knowledge. Mathematics, in particular, plays a central role in numerous professions and has historically served as the gatekeeper to numerous other areas of study, particularly the hard sciences, engineering and business. Mathematics is also a major component in standardized tests in the United States, and in university entrance exams in numerous parts of world.

Creativity and imagination is often evident when young children begin to develop numeric and spatial concepts, and explore mathematical tasks that capture their interest. Creativity is also an essential ingredient in the work of professional mathematicians. Yet, the bulk of mathematical thinking encouraged in the institutionalized setting of schools is focused on rote learning, memorization, and the mastery of numerous skills to solve specific problems prescribed by the curricula or aimed at standardized testing.

Creativity, Giftedness, and Talent Development in Mathematics, pages ix–xi
Copyright © 2008 by Information Age Publishing

My foray into creativity and giftedness began as a school district coordinator of the gifted program in a public school district in Illinois. During this time I was mentored by Professor Robert Wheeler at Northern Illinois University, who shared my interest in the construct of creativity. Several seminars led us into the considerable body of literature on creativity and I naturally wanted to empirically test things out in the classroom. In the public school setting, I tried out many innovative things like integrating science, philosophy and literature with mathematics; conducting teaching experiments with problems that were isomorphic in structure and studying whether students were able to discover generalizations via this process, as well as studies aimed at the insights of mathematicians and gifted students on proof and the nature of creativity in mathematics. Several of the chapters in this book are based on critically peer-reviewed published articles arising from these studies. I also had the good fortune of having the support of Harry Adrian, a teacher of philosophy and great ideas who helped me learn the logistical and practical aspects of operating a functional and equitable gifted program in a public school district. My scholarly interest in the field also grew as a result of serendipity—i.e., meeting Paula Olszweski-Kubilius at the State of Illinois gifted conference at Pheasant Run, St Charles, Illinois in 2000. Paula encouraged me to read the research literature in journals and to start writing and publishing the findings of my research in gifted education journals.

It is difficult to believe that eight years later, many of the research studies that I initiated and published on are receiving wide citations and are also being extended and carried out by scholars in other countries. The chapter by Supattra Pativisan which examines the problem solving abilities of Thai gifted students (Chapter 8) is an extension and a more detailed examination of the ideas found in the chapter entitled *Mathematical giftedness, problem-solving and the ability to formulate generalizations* (Chapter 2). Similarly Chapter 11 by Yim, Song, and Kim on *Mathematically gifted Korean elementary students' revisiting of Euler's polyhedron theorem*, is a practical study inspired by my (audacious) theorization of the possibilities of implementing Lakatosian methods in the classroom (in Chapter 10) based on my results of classroom based studies.

The monograph also contains chapters from Viktor Freiman and Alexandar Karp, scholars who have an insiders perspective on the models of talent development used in the former Soviet Union. In addition the chapters by Sylvia Bulgar, Alan Zollman and Linda Sheffield present a complementary discussion of the issues surrounding mathematics gifted education in the United States. The sheer range of perspectives presented in the chapters and the geographic diversity of the author's backgrounds makes the monograph truly international in its scope.

Given the lack of research based perspectives on talent development in mathematics education, this monograph is specifically focused on contributions towards the constructs of creativity and giftedness in mathematics. This monograph presents new perspectives for talent development in the mathematics classroom and gives insights into the psychology of creativity and giftedness. The book is aimed at classroom teachers, coordinators of gifted programs, math contest coaches, graduate students and researchers interested in creativity, giftedness, and talent development in mathematics.

CHAPTER 1

THE CHARACTERISTICS OF MATHEMATICAL CREATIVITY

Bharath Sriraman
The University of Montana

ABSTRACT

Mathematical creativity ensures the growth of mathematics as a whole. However the source of this growth, the creativity of the mathematician is a relatively unexplored area in mathematics and mathematics education. In order to investigate how mathematicians create mathematics; a qualitative study involving five creative mathematicians was conducted. The mathematicians in this study verbally reflected on the thought processes involved in creating mathematics. Analytic induction was used to analyze the qualitative data in the interview transcripts and to verify the theory driven hypotheses. The results indicate that in general, the mathematicians' creative process followed the four-stage Gestalt model of preparation-incubation-illumination-verification. It was found that social interaction, imagery, heuristics, intuition, and proof were the common characteristics of mathematical creativity. In addition contemporary models of creativity from psychology were reviewed and used to interpret the characteristics of mathematical creativity.

Creativity, Giftedness, and Talent Development in Mathematics, pages 1–31
Copyright © 2008 by Information Age Publishing

1

INTRODUCTION

Mathematical creativity ensures the growth of the field of mathematics as a whole. The constant increase in the number of journals devoted to mathematical research bears evidence to the growth of mathematics. Yet, what lies at the essence of this growth, viz., the creativity of the mathematician has not been the subject of much research. It is usually the case that most mathematicians are uninterested in analyzing the thought processes that result in mathematical creation (Ervynck, 1991). The earliest known attempt to study mathematical creativity was an extensive questionnaire published in the French periodical L'Enseigement Mathematique (1902). This questionnaire and a lecture by the renowned 20th century mathematician Henri Poincaré, to the Societé de Psychologie on creativity inspired his colleague Jacques Hadamard, another prominent 20th century mathematician to investigate the psychology of mathematical creativity (Hadamard, 1945). Hadamard (1945) undertook an informal inquiry among prominent mathematicians and scientists in America such as George Birkhoff, George Polya, and Albert Einstein, about the mental images used in doing mathematics. Hadamard (1945) influenced by Gestalt psychology of his time theorized that mathematicians' creative process followed the four-stage Gestalt model (Wallas, 1926) of preparation-incubation-illumination-verification. The four-stage Gestalt model is a characterization of the mathematician's creative process, which does not define creativity per se. How does one define creativity? In particular what exactly is mathematical creativity? Is it the discovery of a new theorem by a research mathematician? Does student discovery of a hitherto known result also constitute creativity? These are the areas of exploration in this paper.

THE PROBLEM OF DEFINING CREATIVITY

Mathematical creativity has been simply described as discernment, choice (Poincaré, 1948). According to Poincaré (1948), to create consists precisely in not making useless combinations and in making those which are useful and which are only a small minority. This may seem like a vague characterization of mathematical creativity. One can interpret Poincaré's "choice" metaphor to mean the ability of the mathematician to choose carefully between questions (or problems) that bear fruition, as opposed to those that lead to nothing new. But this interpretation does not resolve the fact that Poincaré's definition of creativity overlooks the problem of novelty. In other words, characterizing mathematical creativity as the ability to choose between useful and useless combinations is akin to characterizing the art of sculpting as a process of cutting away the unnecessary!

Poincaré 's (1948) definition of creativity was a result of the circumstances under which he stumbled upon deep results in Fuchsian functions. The first stage consisted of working hard to get an insight into the problem at hand. Poincaré (1948) called this the preliminary period of conscious work. This is also referred to as the *preparatory stage* (Hadamard, 1945). The second stage is when the problem is put aside for a period of time and the mind is occupied with other problems. This is referred to as the *incubatory stage* (Hadamard, 1945). The third stage is where the solution suddenly appears while perhaps engaged in other unrelated activities. "This appearance of sudden illumination is a manifest sign of long, unconscious prior work" (Poincaré, 1948). Hadamard (1945) referred to this as the *illuminatory stage*. However, the creative process does not end here. There is a fourth and final stage, which consists of expressing the results by language or writing. At this stage one verifies the result, makes it precise, and looks for possible extensions through utilization of the result. The Gestalt model has some shortcomings. First, the model mainly applies to problems that have been posed *a priori* by mathematicians, thereby ignoring the fascinating process by which the actual questions are arrived at. Secondly, the model attributes a large portion of what "happens" in the *incubatory-illuminatory* phases to subconscious drives. The problem of how questions are arrived at is partially addressed by Ervynck (1991) in his three-stage model.

Ervynck (1991) described mathematical creativity in terms of three stages. The first stage (*Stage 0*) is referred to as the *preliminary technical stage*, which consists of " some kind of technical or practical application of mathematical rules and procedures, without the user having any awareness of the theoretical foundation." (p. 42). The second stage (*Stage 1*) is that of *algorithmic activity*, which consists primarily of performing mathematical techniques, such as explicitly applying an algorithm repeatedly. The third stage (*Stage 2*) is referred to that as *creative (conceptual, constructive) activity*. This is the stage at which true mathematical creativity occurs and consists of non-algorithmic decision making. "The decisions that have to be taken may be of a widely divergent nature and always involve a choice." (p. 43). Although Ervynck (1991) tries to describe the process by which a mathematician arrives at the questions through his characterizations of *Stage 0* and *Stage 1*, his description of mathematical creativity is very similar to that of Poincaré and Hadamard. In particular his use of the term "non-algorithmic decision making" is analogous to Poincaré's use of the "choice" metaphor.

The author is unaware of any literature in mathematics education that attempts to explicitly define creativity. There are references made to creativity by the soviet researcher Kruteskii (1976) in the context of student's abilities to abstract and generalize mathematical content. There is however one outstanding example of a mathematician (George Polya) attempting to give heuristics to tackle problems in a manner akin to trained mathematicians. Polya (1954) observed that in "trying to solve a problem, we consider different aspects of it in turn, we roll it over and over in our minds; variation of the problem is essential to our work." Polya (1954) emphasized the use of a variety of heuristics for solving mathematical problems of varying complexity. In examining the plausibility of a mathematical conjecture, mathematicians use a variety of strategies. In looking for conspicuous patterns, mathematicians use a variety of heuristics such as (a) verifying consequences, (b) successively verifying several consequences, (c) verifying an improbable consequence, (d) inferring from analogy, (e) deepening the analogy. Thus, heuristics can be viewed as a decision-making mechanism, which lead the mathematician down a certain path, the outcome of which may or may not be fruitful.

As is evident in the preceding paragraphs, the problem of defining creativity is by no means an easy one.

However psychologists renewed interest in the phenomenon of creativity has resulted in literature that attempts to define and operationalize the word "creativity." Recently psychologists have attempted to link creativity to measures of intelligence (Sternberg, 1985), the ability to abstract and generalize (Sternberg, 1985), and to complex problem-solving abilities (Frensch & Sternberg, 1992). Sternberg and Lubart (2000) define creativity as the ability to produce unexpected original work, which is useful and adaptive. Mathematicians would raise several arguments with this definition, simply because the results of creative work may not always have implications that are "useful" in terms of applicability in the real world. A recent example that comes to mind is Andrew Wiles' proof of Fermat's Last Theorem. The mathematical community views his work as creative. It was unexpected and original but had no applicability in the sense of what Sternberg and Lubart (2000) suggest. Hence, I think it is sufficient to define creativity as the ability to produce novel or original work, which is compatible with my personal definition of mathematical creativity as the process that results in unusual and insightful solutions to a given problem, irrespective of the level. In the context of this study involving professional mathematicians, mathematical creativity is defined as the publishing of original results in prominent mathematics research journals.

THE MOTIVATION FOR STUDYING CREATIVITY

The lack of recent mathematics education literature on creativity was one of the motivations for conducting this study. Fifteen years ago, Muir (1988) invited mathematicians to complete a modified version of the original survey in the L'Enseigement Mathematique (1902). The results of this endeavor are of great interest but as yet unknown. The purpose of this study was to gain an insight into the nature of mathematical creativity. This was done by interviewing five accomplished and creative mathematicians, using a modification of the interview protocol that appeared in L'Enseigement Mathematique (1902) and Muir (1988). The purpose of using a modified form of this antiquated questionnaire is discussed in the methodology section of the paper. The author was interested in distilling common attributes of the creative process, to see if there were any underlying themes that characterized mathematical creativity. The specific questions of exploration were:

- Is the Gestalt model of mathematical creativity still applicable today?
- What are the characteristics of the creative process in mathematics?
- Does the study of mathematical creativity have any implications for the classroom?

LITERATURE REVIEW

Any study on the nature of mathematical creativity begs the question as to whether the mathematician discovers or invents mathematics. Therefore the review begins with a brief description of the four popular viewpoints on the nature of mathematics. This is followed by a comprehensive review of contemporary models of creativity from psychology.

The Nature of Mathematics

Mathematicians that are actively involved in research have certain beliefs about the ontological status of mathematics, which influences their approach to research (Davis & Hersh, 1981; Sriraman 2004). The Platonist viewpoint is that mathematical objects exist prior to their discovery, and that "any meaningful question about a mathematical object has a definite answer, whether we are able to determine it or not" (Davis & Hersh, 1981). According to this view, mathematicians do not invent or create mathematics—they discover mathematics. Logicists hold that "all concepts of mathematics can ultimately be reduced to logical concepts" which implies that "all mathematical truths can be proved from the axioms and rules of inference and logic alone," (Ernest, 1991). Formalists do not believe that mathematics is discovered; they believe mathematics is simply a game, created by mathematicians, based on strings of symbols which have no meaning (Davis & Hersh, 1981). Constructivism (incorporating Intuitionism) was one of the major schools of thought (besides Platonism, Logicism and Formalism) that arose due to the contradictions that emerged in the theory of sets and the theory of functions during the early part of the 20th century. Contradictions like Russell's Paradox were a major blow to the absolutist view of mathematical knowledge, for if mathematics is certain and all its theorems are certain, how can contradictions be among its theorems? The early constructivists in mathematics were the intuitionists Brouwer and Heyting. Constructivists claim that both mathematical truths and the existence of mathematical objects must be established by constructivist methods. The constructivist (intuitionist) viewpoint is that "human mathematical activity is fundamental in the creation of new knowledge and that both mathematical truths and the existence of mathematical objects must be established by constructive methods"(Ernest, 1991, p. 29).

The question then is how does a mathematician go about conducting mathematics research? Do the questions appear out of the blue or is there a mode of thinking or inquiry that leads to the meaningful questions and to the methodology for tackling these questions? The author contends that the types of questions asked are determined to a large extent by the culture the mathematician finds herself in. Simply put, it is impossible for an individual to acquire knowledge of the external world without social interaction. According to Ernest (1994) there is no underlying metaphor for the wholly isolated individual mind. Instead the underlying metaphor is that of persons in conversation, comprising persons in meaningful linguistic interaction and dialogue (Ernest, 1994). Language is the shaper of, as well as being the "summative" product of individual minds (Wittgenstein, 1978). The recent literature in psychology acknowledges these social dimensions of human activity as being instrumental in the creative process. This warrants an in-depth review of this literature.

The Notion of Creativity in Psychology

As stated earlier, research on creativity has been on the fringes of psychology, educational psychology and mathematics education. It is only in the last twenty-five years that there has been a renewed interest in the phenomenon of creativity in the psychology community. *The Handbook of Creativity* (Sternberg, 2000) which contains a comprehensive review of all research available in the field of creativity suggests that most of the approaches used in the study of creativity can be subsumed under six categories, namely: mystical, pragmatic, psychodynamic, psychometric, cognitive and social-personality. Each of these approaches are briefly reviewed.

The Mystical Approach
The mystical approach to studying creativity suggests that creativity is the result of divine inspiration, or is a spiritual process. In the history of mathematics, Blaise Pascal claimed that many of his mathematical insights

came directly from God. The renowned 19th century algebraist Leopold Kronecker said that "God made the integers, all the rest is the work of man" (Gallian, 1994). Although his radical beliefs did not attract many supporters, the intuitionists advocated his beliefs about constructive proofs many years after his death. There have been attempts to explore possible relationships between the mathematician's belief on the nature of mathematics and their creativity (Davis & Hersh, 1981; Hadamard, 1945; Poincaré, 1948; Sriraman, 2004). These studies indicate that there is certainly a relationship between a mathematician's belief on the nature of mathematics and creativity. It is commonly believed that the neo-Platonist view is helpful to the research mathematician because of the innate belief that the sought after result/relationship already exists.

The Pragmatic Approach

The pragmatic approach entails "being concerned primarily with developing creativity" (Sternberg, 2000, p. 5), as opposed to understanding it. Polya's (1954) emphasis on the use of a variety of heuristics for solving mathematical problems of varying complexity is an example of a pragmatic approach. Thus, heuristics can be viewed as a decision-making mechanism, which lead the mathematician down a certain path, the outcome of which may or may not be fruitful. The popular technique of brainstorming used in corporations is another example of inducing creativity by seeking as many ideas or solutions possible in a non-critical setting.

The Psychodynamic Approach

The psychodynamic approach to studying creativity is based on the idea that creativity arises from the tension between conscious reality and unconscious drives (Hadamard, 1945; Poincaré, 1948, Sternberg, 2000, Wallas, 1926; Wertheimer, 1945). The four-step gestalt model (preparation-incubation-illumination-verification) is an example of the use of a psychodynamic approach to studying creativity. It should be noted that the gestalt model has served as the kindling for many contemporary problem-solving models (Polya, 1945; Schoenfeld ,1985; Lester, 1985). Early psychodynamic approaches to creativity were used to construct case studies of eminent creators such as Albert Einstein, but this approach was criticized by the behaviorists because of the difficulty of measuring proposed theoretical constructs.

The Psychometric Approach

The psychometric approach to studying creativity entails quantifying the notion of creativity with the aid of paper and pencil tasks. An example of this would be the Torrance Tests of Creative Thinking developed by Torrance (1974), which are used by many gifted programs in middle and high schools, to identify students that are gifted/creative. This test consists of several verbal and figural tasks that call for problem-solving skills and divergent thinking. The test is scored for fluency, flexibility, originality (the statistical rarity of a response) and elaboration (Sternberg, 2000). Sternberg (2000) states that there are positive and negative sides to the psychometric approach. On the positive side, these tests allow for research with non-eminent people, are easy to administer, and objectively scored. The negative side is that numerical scores fail to capture the concept of creativity because they are based on brief paper and pencil tests. Researchers call for use of more significant productions such as writing samples, drawings, etc to be subjectively evaluated by a panel of experts instead of simply relying on a numerical measure.

The Cognitive Approach

The cognitive approach to the study of creativity focuses on understanding the "mental representations and processes underlying human thought" (Sternberg, 2000, p. 7). Weisberg (1993) suggests that creativity entails the use of ordinary cognitive processes and results in original and extraordinary products. These products are the result of cognitive processes acting on the knowledge already stored in the memory of the individual. There is a significant amount of literature in the area of information-processing (Birkhoff, 1969; Minsky, 1985) that attempts to isolate and explain cognitive processes in terms of machine metaphors.

The Social-Personality Approach

The social-personality approach to studying creativity focuses on personality and motivational variables as well as the socio-cultural environment as sources of creativity. Sternberg (2000) states that numerous studies conducted at the societal level, indicate that "eminent levels of creativity over large spans of time are statistically linked to variables such as cultural diversity, war, availability of role models, availability of financial support, and competitors in a domain." (p. 9)

Most of the recent literature on creativity (Csikszentmihalyi, 1988, 2000; Gruber & Wallace, 2000; Sternberg & Lubart, 1996) suggests that creativity is the result of confluence of one or more of the factors from these six aforementioned categories. The "confluence" approach to the study of creativity has gained credibility and the research literature has numerous confluence theories for better understanding the process of creativity. This calls for a review of the most commonly cited confluence theories of creativity. This is followed by a description of the methodology employed for data collection and data analysis in this study.

Confluence Theories of Creativity

The three most commonly cited "confluence" approaches to the study of creativity are the "systems approach" (Csikszentmihalyi, 1988, 2000); "the case study as evolving systems approach" (Gruber & Wallace, 2000), and finally the "investment theory approach" (Sternberg & Lubart, 1996).

The Systems Approach

The systems approach takes into account the social and cultural dimensions of creativity, instead of simply viewing creativity as an individualistic psychological process. The system approach studies the interaction between the individual, domain and field. The field consists of people who have influence over a domain. For example, editors of mathematics research journals would have influence on the domain of mathematics. The domain is in a sense a cultural organism that preserves and transmits creative products to other individuals in the field. The systems model suggests that creativity is a process that is observable at the "intersection where individuals, domains and fields interact" (Csikzentmihalyi, 2000). The three components, namely, individual, domain and field are necessary because the individual operates in a cultural or symbolic (domain) aspect as well as a social (field) aspect.

"The domain is a necessary component of creativity because it is impossible to introduce a variation without reference to an existing pattern. New is meaningful only in reference to the old" (Csikszentmihalyi, 2000). Thus creativity occurs when an individual makes a change in a given domain, and this change is transmitted through time. The personal background of an individual and their position in a domain naturally influence the likelihood of their contribution. For example, a mathematician working immersed in the culture of a research university is more likely to produce research papers because of the time available for "thinking" as well as being immersed in a culture where ideas flourish. It is no coincidence that in the history of science, there are significant contributions from clergymen such as Pascal, and Mendel, to name a few, because they had the means and the leisure to "think." Csikszentmihalyi (2000) then argues that novel ideas that result in significant changes are unlikely to be adopted unless they are sanctioned by a group of experts that decide what gets included in the domain. These "gatekeepers" (experts) constitute the field. For example, in mathematics, the opinion of a very small number of leading researchers was enough to certify the validity of Andrew Wiles' proof to Fermat's Last Theorem.

There are numerous examples within the field of mathematics that fall within the systems model. For instance the Bourbaki, a group of mostly French mathematicians who began meeting in the 1930s, aimed to write a thorough unified account of all mathematics. The Bourbaki were essentially a group of expert mathematicians that tried to unify all of mathematics and become the gatekeepers of the field so to speak by setting the standard for rigor. Although the Bourbakists failed in their attempt, students of the Bourbakists, who are editors of certain prominent journals to this day demand a very high degree of rigor in submitted articles, thereby serving as gatekeepers of the field.

A different example is that of the role of proof. Proof is the social process through which the mathematical community validates the mathematician's creative work (Hanna, 1991). The Russian logician Manin (1977) said "A proof becomes a proof after the social act of accepting it as a proof. This is true of mathematics as it is of physics, linguistics, and biology."

In summary, the systems model of creativity suggests that for creativity to occur a set of rules and practices must be transmitted from the domain to the individual. The individual then must produce a novel variation in the content of the domain, and this variation must be selected by the field for inclusion in the domain.

Gruber and Wallace's Case Study as Evolving Systems Approach

In contrast to Csikszentmihalyi's (2000) argument that calls for focus on communities in which creativity manifests, Gruber and Wallace (2000) propose a model that treats each individual as a unique evolving system of creativity and ideas, and therefore each individual's creative work must be studied on their own. This viewpoint of Gruber and Wallace (2000) is a belated victory of sorts for the Gestaltists, who essentially proclaimed the same thing almost a century ago. Gruber and Wallace's (2000) use of terminology that jives with current trends in psychology seems to make their ideas more acceptable. They propose a model that calls for "detailed analytic and sometimes narrative descriptions of each case and efforts to understand each case as a unique functioning system (Gruber & Wallace, 2000, p. 93). It is important to note that the emphasis of this model is not to explain the origins of creativity, nor is it the personality of the creative individual, but on "how creative work works?" (p. 94). The questions of concern to Gruber and Wallace are: 1) What do creative people do when they are being creative? and 2) how do creative people deploy available resources to accomplish something unique? In this model creative work is defined as one that is novel and has value. This definition is consistent with that used by current researchers in creativity (Csikszentmihalyi, 2000; Sternberg & Lubart, 2000). Gruber and Wallace (2000) also claim that creative work is always the result of purposeful behavior and that creative work is usually a long undertaking "reckoned in months, years and decades"(p. 94). The author does not agree with the claim that creative work is always the result of purposeful behavior. One counterexample that comes to mind is the discovery of penicillin. The discovery of penicillin could be attributed purely to chance. On the other hand there are numerous examples that support the claim that creative work sometimes entails work that spans years. In mathematical folklore there are numerous examples of such creative work. For example, Kepler's laws of planetary motion were the result of twenty years of numerical calculations. Andrew Wiles' proof of Fermat's Last Theorem was a seven year undertaking. The Riemann hypothesis states that the roots of the zeta function (complex numbers z, at which the zeta function equals zero) lie on the line parallel to the imaginary axis and half a unit to the right of it. This is perhaps the most outstanding unsolved problem in mathematics with numerous implications. The analyst Levinson undertook a determined calculation on his deathbed that increased the credibility of the Riemann-hypothesis. This is another example of creative work that falls within Gruber and Wallace's (2000) model.

The case study as an evolving system has the following components to it. First, it views creative work as multi-faceted. So, in constructing a case study of a creative work, one has to distill out the facets that are relevant and construct the case study based on the chosen facets. Some facets that can be used to construct an evolving system case study are (a) Uniqueness of the work; (b) a narrative of what the creator achieved; (c) systems of belief; (d) multiple time-scales (construct the time-scales involved in the production of the creative work); (e) problem solving; and (f) contextual frame (family, schooling, teacher's influences) (Gruber & Wallace, 2000). So in summary, constructing a case study of a creative work as an evolving system entails incorporating the many facets suggested by Gruber and Wallace (2000). One could also evaluate a case study involving creative work by looking for the above mentioned facets.

The Investment Theory Approach

According to the investment theory model, creative people are like good investors, that is, they buy low and sell high (Sternberg & Lubart, 1996). The context here is naturally in the realm of ideas. Creative people conjure up ideas that are either unpopular or disrespected, but invest considerable time convincing other people about the intrinsic worth of these ideas (Sternberg & Lubart, 1996). They sell high in the sense that they let other people pursue their ideas while they move on to the next idea. Investment theory claims that the convergence of six elements constitutes creativity. The six elements are intelligence, knowledge, thinking styles, personality, motivation and environment. It is also important that the reader not mistake the word intelligence to an IQ score. On the contrary Sternberg (1985) suggests a triarchic theory of intelligence which consists of synthetic (ability to generate novel, task appropriate ideas), analytic and practical abilities. Knowledge is defined as knowing enough about a particular field to move it forward. Thinking styles are defined as a preference for thinking

in original ways of one's choosing, the ability to think globally as well as lo-
cally and the ability to distinguish questions of importance from those that
are not important. Personality attributes that foster creative functioning
are the willingness to take risks, overcome obstacles and tolerate ambiguity.
Finally motivation and an environment that is supportive and rewarding are
also essential elements of creativity (Sternberg, 1985).

In investment theory creativity involves the interaction between a person, task, and environment. This is in a sense a particular case of the systems model (Csikszentmihalyi, 2000). The implication of viewing creativity
as the interaction between person, task and environment is that what is
considered novel or original may vary from one person, task, and environment to another. The investment theory model suggests that creativity is
more than a simple sum of the attained level of functioning in each of
the six elements. Regardless of the functioning levels in other elements, a
certain level or threshold of knowledge is required without which creativity
is impossible. High levels of intelligence and motivation can positively enhance creativity, and compensations can occur to counteract weaknesses in
other elements. For example, one could be in an environment that is non-
supportive of creative efforts, but a high level of motivation would possibly
overcome this and pursue creative endeavors.

This concludes the review of three commonly cited prototypical confluence theories of creativity, namely the systems approach (Csikszentmihalyi,
2000), which suggests that creativity is a sociocultural process involving the
interaction between the individual, domain and field; Gruber and Wallace's (2000) model that treats each individual case study as a unique evolving system of creativity; and investment theory (Sternberg & Lubart,1996),
which suggests that creativity is the result of the convergence of six elements (intelligence, knowledge, thinking styles, personality, motivation
and environment).

Having reviewed the research literature on creativity, the focus is shifted
on the methodology employed for studying mathematical creativity.

METHODOLOGY

The Interview Instrument

The purpose of this study was to gain an insight into the nature of mathematical creativity. The author was interested in distilling common attributes from how mathematicians created mathematics, in order to determine some of the characteristics of the creative process. The author was
also interested in testing the applicability of the gestalt model. The primary method of data collection was through personal interviews. Since the

main focus of the study was to ascertain qualitative aspects of creativity, a formal interview methodology was selected. The interview instrument (Appendix A) was developed by modifying questions from questionnaires in L'Enseigement Mathematique (1902) and Muir (1988). The purpose of using this modified questionnaire was as follows. The questions were general in nature, which allowed the mathematicians to express themselves freely. Secondly the author wanted to somehow test the applicability of the four stage Gestalt model of creativity. Therefore the existing instruments were modified in order to operationalize the Gestalt theory and to allow the natural flow of ideas to form the basis of a thesis that would emerge from this exploration.

Background of the Subjects

Five mathematicians from the mathematical sciences faculty at a large Ph.D. granting mid-western university were selected. These mathematicians were chosen based on their accomplishments and the diversity of the mathematical areas they worked in. This was measured by counting the number of published papers in prominent research journals, as well as the diversity of the mathematical domains in which the mathematicians conducted research. Four of the mathematicians were tenured full professors, who had been professional mathematicians for over 30 years. Only one of the mathematicians was considerably younger, and a tenured associate professor. All the interviews were conducted formally in a closed door setting, in the mathematicians' office. The interviews were taped and transcribed verbatim.

Data Analysis

Since creativity is an extremely complex construct involving a wide range of interacting behaviors, in the author's opinion it should be studied holistically. The principle of analytic induction (Patton, 2002) was applied to the interview transcripts to discover dominant themes, which described the behavior under study. According to Patton (2002) "analytic induction, in contrast to grounded theory, begins with an analyst's deduced propositions or theory-derived hypotheses and 'is a procedure for verifying theories and propositions based on qualitative data' (Taylor & Bogdan, 1984, p. 127)." Following the principles of analytic induction, the data was carefully analyzed in order to extract common strands. These common strands were then compared to theoretical constructs in the existing literature with the explicit purposes of verifying whether the Gestalt model was applicable to this qualitative data, as well as to extract themes that characterized the mathematician's creative process. If an emerging theme could not be classified or named because of being unable to grasp its properties or significance, then theoretical comparisons were made. Corbin and Strauss (1998) state that "using comparisons brings out properties, which in turn can be used to examine the incident or object in the data. The specific incidents, objects, or actions that we use when making theoretical comparisons can be derived from the literature and experience. It is not that we use experience or literature as data *but rather that we use the properties and dimensions derived from the comparative incidents to examine the data in front of us.*" (p. 80). Themes that emerged were social interaction, preparation, use of heuristics, imagery, incubation, illumination, verification, intuition and proof.

Vignettes that highlight these characteristics are reconstructed in the next section along with commentaries that incorporate the wider conversation, and a continuous discussion of connections to the existing literature.

RESULTS, COMMENTARIES AND DISCUSSION

All the mathematicians in this study were tenured professors at a large Ph.D. granting mathematical sciences faculty. Their place of work could be described as academic, with teaching and committee duties. The mathematicians were free to choose their areas of research and the problems that they worked. Four of the five mathematicians had worked and published as individuals with occasional joint ventures. Only one of the mathematicians had done extensive collaborative work. All but one of the mathematicians did not structure their time for research mathematics. The main reasons cited were family commitments and teaching responsibilities during the regular school year. All the mathematicians found it easier to concentrate on research in the summers because of lighter or non-existent teaching responsibilities. Two of the mathematicians showed a pre-disposition towards mathematics at the early secondary school level. The others became interested in mathematics only at a later stage in their university education. None of the mathematicians who participated in this study had any immediate family influence that was of primary importance in their mathematical development. Four of the mathematicians recalled being influenced by particular teachers at various stages of their education. One of the mathematicians was influenced by a textbook. The three mathematicians who worked primarily in analysis strived consciously to obtain a broad overview of mathematics not of immediate relevance to their main interests. The two algebraists expressed interest in other areas of mathematics but were active mainly in their chosen fields.

Supervision of Research and Social Interaction

As reported earlier, all the mathematicians in this study were tenured professors in a research university. Besides teaching, research, and committee obligations, many mathematicians play a big role in mentoring graduate students interested in their area of research. Research supervision is an aspect of creativity because any interaction between human beings is the ideal setting for exchange of ideas. During this interaction the mathematician is exposed to different perspectives on the subject. All the mathematicians in this study valued the interaction they had with their graduate students. Vignettes of individual responses follow.[1]

Vignette 1

A: I've had only one graduate student per se and she is just finishing up her Ph.D right now and I'd say it has been a very good interaction to see somebody else get interested in the subject and come up with new ideas, and exploring those ideas with her.

B: I have had a couple of students who have sort of started but who haven't continued on to a Ph.D, so I really can't speak to that. But the interaction was positive.

C: Of course, I have a lot of collaborators, these are my former students you know... I am always all the time working with students, this is normal situation.

D: That is difficult to answer (silence)... it is positive because it is good to interact with other people. It is negative because it can take a lot of time. As you get older your brain doesn't work as well as it used to and... younger people by and large their minds are more open, there is less garbage in there already. So, it is exciting to work with younger people who are in their most creative time. When you are older, you have more experience, when you are younger your mind works faster... not as fettered.

E: Oh... it is a positive factor I think, because it continues to stimulate ideas... talking about things and it also reviews things for you in the process, puts things in perspective, and keep the big picture. It is helpful really in your own research to supervise students.

Commentary on Vignette 1

The responses of the mathematicians in the preceding vignette are focussed on research supervision. However all the mathematicians acknowledged the role of social interaction in general as an important aspect that stimulated creative work. Many of the mathematicians mentioned the advantages of being able to e-mail colleagues, going to research conferences and other professional meetings. This is revealed more in the following section under preparation.

Preparation and the Use of Heuristics

When mathematicians are about to investigate a new topic, there is usually a body of existing research in the area of the new topic. One of goals of the study was to find out how creative mathematicians approached a new topic or a problem. Did they try their own approach, or did they first attempt to assimilate what was already known about that topic? Did the mathematicians make use of computers to gain insight into the problem? What were the various modes of approaching a new topic or problem? The responses indicate that a variety of approaches were used.

Vignette 2

A: Talk to people who have been doing this topic. Learn the types of questions that come up. Then I do basic research on the main ideas. I find that talking to people helps a lot more than reading because you get more of a feel for what the motivation is beneath everything.

B: What might happen for me, is that I may start reading something, and, if feel I can do a better job, then I would strike off on my own. But for the most part I would like to not have to re-invent a lot that is already there. So, a lot of what has motivated my research has been the desire to understand an area. So, if somebody has already laid the groundwork then it's helpful. Still I think a large part of doing research is to read the work that other people have done.

C: It is connected with one thing that simply... my style was that I worked very much and I even work when I could not work. Simply the problems that I solve attract me so much, that the question was who will die first... mathematics or me? It was never clear who would die.

D: Try and find out what is known. I won't say assimilate... try and find out what's known and get an overview, and try and let the problem speak... mostly by reading because you don't have that much immediate contact with other people in the field. But I find that I get more from listening to talks that other people are giving than reading.

> **E:** Well! I have been taught to be a good scholar. A good scholar attempts to find out what is first known about something or other before they spend their time simply going it on their own. That doesn't mean that I don't simultaneously try to work on something.

Commentary on Vignette 2

These responses indicate that the mathematician spends a considerable amount of time researching the context of the problem. This is primarily done by reading the existing literature and by talking to other mathematicians in the new area. This finding is consistent with the systems model, which suggests that creativity is a dynamic process involving the interaction between the individual, domain and field (Csikzentmihalyi, 2000).

At this stage it is reasonable to ask whether a mathematician works on a single problem until a breakthrough occurs or do mathematicians work on several problems concurrently? It was found that each of the mathematicians worked on several problems concurrently, using a back and forth approach.

Vignette 3

> **A:** I work on several different problems for a protracted period of time ... there have been times when I have felt, yes, I should be able to prove this result, then I would concentrate on that thing for a while but they tend to be several different things that I thinking about a particular stage.
>
> **B:** I probably tend to work on several problems at the same time. There are several different questions that I am working on ... mmm ... probably the real question is how often do you change the focus? Do I work on two different problems on the same day? And that is probably up to whatever comes to mind in that particular time frame. I might start working on one rather than the other. But I would tend to focus on one particular problem for a period of weeks, then you switch to something else. Probably what happens is that I work on something and I reach a dead end then I may shift gears and work on a different problem for a while, reach a dead end there and come back to the original problem, so its back and forth.
>
> **C:** I must simply think on one thing and not switch so much.
>
> **D:** I find that I probably work on one. There might be a couple of things floating around but I am working on one

and if I am not getting anywhere, then I might work on the other and then go back.

E: I usually have couple of things going. When I get stale on one, then I will pick up the other, and bounce back and forth. Usually I have one that is primarily my focus at a given time and I will spend time on it over another, but it is not uncommon for me to have a couple of problems going at a given time. Sometimes when I am looking for an example that is not coming, instead of spending my time beating my head against the wall, looking for that example is not a very good use of time. Working of another helps to generate ideas that I can bring back to the other problem.

Commentary on Vignette 3

The preceding vignette indicates that mathematicians tend to work on more than one problem at a given

time. Do mathematicians switch back and forth between problems in a completely random manner, or do they employ and exhaust a systematic train of thought about a problem before switching to a different problem? Many of the mathematicians reported using heuristic reasoning, trying to prove something one day and disprove it the next day, looking for both examples and counterexamples, the use of "manipulations" (Polya, 1954) to gain an insight into the problem. This indicates that mathematicians do employ some of the heuristics made explicit by Polya. It was unclear whether the mathematicians made use of computers to gain an experimental or computational insight into the problem. The author was also interested in knowing the types of imagery used by mathematicians in their work. The mathematicians in this study were queried about this and the following vignette gives us an insight into this aspect of mathematical creativity.

Imagery

The mathematicians in this study were asked about the kinds of imagery that they used to think about mathematical objects. This is reported here with the hope that the reader gets a glimpse into how mathematicians think of mathematical objects. The responses also point out the difficulty of explicitly describing imagery.

Vignette 4

A: Yes I do, yes I do, I tend to draw a lot of pictures when I am doing research, I tend to manipulate things in the air, you know to try to figure out how things work. I have a very geometrically based intuition and uhh . . . so very definitely I do a lot of manipulations.

B: That is a problem because of the particular area I am in. I can't draw any diagrams, things are infinite, so I would love to be able to get some kind of a computer diagram to show the complexity for a particular ring . . . to have something like the Julia sets or . . . mmm . . . fractal images, things which are infinite but you can focus in closer and closer to see possible relationships. I have thought about that with possibilities on the computer. To think about the most basic ring, you would have to think of the ring of integers and all of the relationships for divisibility, so how do you somehow describe this tree of divisibility for integers . . . it is infinite.

C: Science is language, *you think through language.* But it is language simply; you put together theorems by logic. You first see the theorem in nature . . . you must see that somewhat is reasonable and then you go and begin and then of course there is big, big, big work to just come to some theorem in non-linear elliptic equations . . .

D: A lot of mathematics, whether we are teaching or doing, is attaching meaning to what we are doing and this is going back to the earlier question when you talked about how do you do it, what kind of heuristics do you use? What kind of images do you have that you are using. A lot of doing mathematics is creating these abstract images that connect things and then making sense of them but that doesn't appear in proofs either.

E: Pictorial, linguistic, kinesthetic . . . any of them is the point right! Sometimes you think of one, sometimes another. It really depends on the problem you are looking at, they are very much . . . often I think of functions as very kinesthetic, moving things from here to there. Other approaches you are talking about is going to vary from problem to problem, or even day to day. Sometimes when I am working on research, I try to view things in a s many different ways as possible, to see what is really happening. So there are a variety of approaches.

Commentary on Vignette 4

Besides revealing the difficulty of describing mental imagery, all the mathematicians reported that they did not use computers in their work.

This characteristic of the pure mathematician's work is echoed in Poincaré's (1948) use of the "*choice*" metaphor and Ervynck's (1991) use of the term "*non-algorithmic decision making.*" The doubts expressed by the mathematicians about the incapability of machines to do their work brings to mind the reported words of Garrett Birkhoff, one of the great, applied mathematicians of our time. In his retiring presidential address to the SIAM, Birkhoff (1969) addressed the role of machines in human creative endeavors. In particular part of this address was devoted to discussing the psychology of the mathematicians (and hence of mathematics). Birkhoff said:

> The remarkable recent achievements of computers have partially fulfilled an old dream. These achievements have led some people to speculate that tomorrow's computers will be even more "intelligent" than humans, especially in their powers of mathematical reasoning...the ability of good mathematicians to sense the significant and to avoid undue repetition seems however hard to computerize; without it, the computer has to pursue millions of fruitless paths avoided by experienced human mathematicians. (Birkhoff, 1969, pp. 430–438)

Incubation and Illumination

Having reported on the role of research supervision and social interaction, the use of heuristics and imagery, which can be viewed as aspects of the preparatory stage of mathematical creativity, it is natural to ask what occurs next. As the literature theorizes, after the mathematician works hard to gain insight into a problem, there is usually a transition period (conscious work on the problem ceases and unconscious work begins), where the problem is put aside before the breakthrough occurs. The mathematicians in this study reported experiences, which are consistent with the existing literature (Hadamard, 1945; Poincaré, 1948).

Vignette 5

> **B:** One of the problems is first one does some preparatory work, that has to be the left side, and then you let it sit. I don't think you get ideas out of nowhere, you have to do the groundwork first, okay. This is why people will say, now we have worked on this problem, so let us sleep on it. So you do the preparation, so that the sub-conscious or intuitive side may work on it and the answer comes back but you can't really tell when. You have to be open to this, lay the groundwork, think about it and then these flashes

of intuition come and they represent the other side of the brain communicating with you at whatever odd time.

D: I am not sure you can really separate them because they are somewhat connected. You spend a lot of time working on something and you are not getting anywhere with it...with the deliberate effort, then I think your mind continues to work and organize. And maybe when the pressure is off the idea comes...but the idea comes because of the hard work.

E: Usually they come after I have worked very hard on something or another, but they may come at an odd moment. They may come into my head before I go bed....What do I do at that point? Yes I write it down (laughing). Sometimes when I am walking somewhere, the mind flows back to it (the problem) and says what about that, why don't you try that. That sort of thing happens. One of the best ideas I had was when I was working on my thesis....Saturday night, having worked on it quite a bit, sitting back and saying why don't I think about it again...and ping! There it was...I knew what it was, I could do that. Often ideas are handed to you from the outside, but they don't come until you have worked on it long enough.

Commentary on Vignette 5

As is evident in the preceding vignette, three out of the five mathematicians reported experiences consistent with the Gestalt model. Mathematician C attributed his breakthroughs on problems to his unflinching will of never giving up and to divine inspiration, echoing the voice of Pascal in a sense. However Mathematician A attributed breakthroughs to chance. In other words making the appropriate (psychological) connections by pure chance which eventually result in the sought after result. The author thinks that it is necessary to comment about the unusual view of mathematician A. Chance plays an important role in mathematical creativity. Great ideas and insights may be the result of chance such as the discovery of penicillin. Ulam (1976) estimated that there is a yearly output of 200,000 theorems in mathematics. Chance plays a role in what is considered important in mathematical research since only a handful of results and techniques survive out of the volumes of published research. The author wishes to draw a distinction between chance in the "Darwinian" sense (as to what survives), and chance in the psychological sense (which results in discovery/invention). The role of chance is addressed by Muir (1988) as follows:

> The act of creation of new entities has two aspects: the generation of new possibilities, for which we might attempt a stochastic description, and the selection of what is valuable from among them. However the importation of biological metaphors to explain cultural evolution is dubious...both creation and selection are acts of design within a social context. (Muir, 1988 p. 33)

Thus, Muir (1988) rejects the Darwinian explanation. On the other hand, Nicolle (1932) in *Biologie de L'Invention* does not acknowledge the role of unconsciously present prior work in the creative process. He attributes breakthroughs to pure chance:

> By a streak of lightning, the hitherto obscure problem, which no ordinary feeble lamp would have revealed, is at once flooded in light. It is like a creation. Contrary to progressive acquirements, such an act owes nothing to logic or to reason. The act of discovery is an accident. (Hadamard, 1945, p. 19)

Nicolle's Darwinian explanation was rejected by Hadamard on the grounds that to explain creation occurs by pure chance is equivalent to asserting that there are effects without causes. Hadamard further argued that although Poincaré attributed his particular breakthrough in Fuchsian functions to chance, Poincaré did acknowledge that there was a considerable amount of previous conscious effort, followed by a period of unconscious work. Hadamard (1945) further argued that even if Poincaré's breakthrough was a result of chance alone, chance alone was insufficient to explain the considerable body of creative work credited to Poincaré in almost every area of mathematics. The question then is how does (psychological) chance work?

The author conjectures that the mind throws out fragments (ideas) which are products of past experience. Some of these random fragments can be juxtaposed and combined in a meaningful way. For example, if one reads a complicated proof consisting of a thousand steps, a thousand random fragments may not be enough to construct a meaningful proof. However the mind chooses relevant fragments from these random fragments and links them into something meaningful. Wedderburn's theorem, that a finite division ring is a field is one instance of a unification of apparently random fragments because the proof involves algebra, complex analysis and number theory.

Polya (1954) addresses the role of chance in a probabilistic sense. It often occurs in mathematics, that a series of mathematical trials (involving computation) throw up numbers which are close to a Platonic ideal. The classic example is Euler's investigation of the infinite series $1 + \frac{1}{4} + \frac{1}{9} + \frac{1}{16} + \ldots + \frac{1}{n^2} + \ldots$ Euler obtained an approximate numerical value for the sum of the series using various transformations of the series. The numerical approximation was 1.644934. Euler confidently guessed the sum of the series to be $\pi^2/6$.

Although the numerical value obtained by Euler and the value of $\pi^2/6$ coincided up to seven decimal places, such a coincidence could be attributed to chance. However, a simple calculation shows that the probability of seven digits coinciding is one in ten million! Hence Euler did not attribute this coincidence to chance and boldly conjectured that the sum of this series was indeed $\pi^2/6$, not to mention the fact that he later proved his conjecture to be true (Polya, 1954, pp. 95–96).

Intuition, Verification and Proof

Once illumination has occurred whether or not through sheer chance or through incubation or through divine intervention, mathematicians usually try to verify that their intuitions were correct, and try to construct a proof. This section of the paper discusses how mathematicians go about verifying their intuitions, and the role of formal proof in the creative process. The mathematicians in this study were asked to describe how they went about forming an intuition about the truth of a proposition. They were asked whether they relied on repeatedly checking a formal proof, or if they used multiple converging partial proofs, or if they first looked for coherence with other results in the area, or if they looked at applications. Most of the mathematicians in this study mentioned that the last thing they looked at was a formal proof. This is consistent with the literature on the role of formal proof in mathematics (Polya, 1954; Usiskin, 1987). Most of the mathematicians mentioned the need for coherence with other results in the area. The mathematician's responses to this question follow.

Vignette 6

> **B:** I think I would go for repeated checking of the formal proof... but I don't think that that is really enough. All of the others have to also be taken into account. I mean, you can believe that something is true although you may not fully understand it. This is the point that was made in the lecture by ___of ___ University on Dirichlet series. He was saying that we have had a formal proof for some time, but that is not to say that it is really understood, and what did he mean by that? Not that the proof wasn't understood, but it was the implications of the result that are not understood, their connections with other results, applications and why things really work? But probably the first thing that I would really want to do is check the formal proof to my satisfaction, so that I believe that it is

correct although at that point I really do not understand its implications... it is safe to say that it is my surest guide.

C: First you must see it in the nature, something, first you must see that this theorem corresponds to something in nature, then if you have this impression, it is something relatively reasonable, then you go to proofs... and of course I have also several theorems and proofs that are wrong, but the major amount of proofs and theorems are right.

D: The last thing that comes is the formal proof. I look for analogies with other things.... How your results that you think might be true would illuminate other things and would fit in the general structure.

E: Since I work in an area of basic research, it is usually coherence with other things, that is probably more than anything else. Yes, one could go back and check the proof and that sort of thing but usually the applications are yet to come, they aren't there already. Usually what guides the choice of the problem is the potential for application, part of what represents good problems is their potential for use. So, you certainly look to see if it makes sense in the big picture... that is a coherence phenomenon. Among those you've given me, that's probably the most that fits.

Commentary on Vignette 6

This vignette indicates that for mathematicians, valid proofs have varied degrees of rigor. "Among mathematicians, rigor varies depending on time and circumstance, and few proofs in mathematics journals meet the criteria used by secondary school geometry teachers (each statement of proof is backed by reasons). Generally one increases rigor only when the result does not seem to be correct" (Usiskin, 1987). Proofs are in most cases the final step in this testing process. "Mathematics in the making resembles any other human knowledge in the making. The result of the mathematician's creative work is demonstrative reasoning, a proof; but the proof is discovered by plausible reasoning, by guessing" (Polya, 1954). How mathematicians approached proof in this study was very different from the logical approach found in proof in most textbooks. The logical approach is an artificial reconstruction of discoveries that are being forced into a deductive system and in this process the intuition that guided the discovery process gets lost.

CONCLUSIONS AND IMPLICATIONS

The goal of this study was to gain an insight into mathematical creativity. As suggested by the literature review, the existing literature on mathematical creativity is relatively sparse. In trying to better understand the process of creativity, the author finds that the Gestalt model proposed by Hadamard (1945) is still applicable today. This study has attempted to add some detail to the preparation-incubation-illumination-verification model of Hadamard (1945), by taking into account the role of imagery, the role of intuition, the role of social interaction, the use of heuristics, and the necessity of proof in the creative process. The mathematicians worked in a setting that was conducive to prolonged research. There was a convergence of intelligence, knowledge, thinking styles, personality, motivation and environment that enabled them to work creatively (Sternberg, 2000; Sternberg & Lubart, 1996, 2000). The preparatory stage of mathematical creativity consisted of various approaches used by the mathematician to lay the groundwork. These are, reading the existing literature, talking to other mathematicians in the particular mathematical domain (Csikzentmihalyi, 1988, 2000), trying a variety of heuristics (Polya, 1954), using a back-and-forth approach of plausible guessing. One of the mathematicians said that he first looked to see if the sought after relationships corresponded to natural phenomenon.

All of the mathematicians in this study worked on more than one problem at a given moment. This is consistent within the investment theory view of creativity (Sternberg & Lubart, 1996). The mathematicians invested an optimal amount of time on a given problem, but switched to a different problem if no breakthrough was forthcoming. All the mathematicians in this study considered this as the most important and difficult stage of creativity. The prolonged hard work was followed by a period of incubation where the problem was put aside (and the preparatory stage is repeated for a different problem). Thus, there was a transition in the mind from conscious to unconscious work on the problem. One mathematician said that this is the stage where the "problem begins to talk to you." Another mathematician said that at this stage the intuitive side of the brain began communicating with the logical side, and conjectured that this communication was not possible at a conscious level.

The transition from incubation to illumination occurred when least expected. Many reported the breakthrough occurring as they were going to bed, or walking, or sometimes a result of speaking to someone else about the problem. One mathematician described this as follows. "You talk to somebody and they say just something that might have been very ordinary a month before but if they say it when you are ready for it, and Oh yeah, I can do it that way can't I! But you have to be ready for it. Opportunity knocks but you have to be able to answer the door."

Illumination follows with the mathematician verifying the occurred idea by constructing a proof. In this study, most of the mathematicians looked for coherence of the result with other existing results in the area of research. If the result cohered with other results, and fit the general structure of the area, only then did the mathematician try to construct a formal proof. In terms of the mathematician's beliefs about the nature of mathematics and its influence on their research the study revealed that four of the mathematicians leaned towards platonism, running contrary to the popular notion that platonism is an exception today. A detailed discussion of this aspect of the research is beyond the scope of this paper. However it was found that beliefs regarding the nature of mathematics influenced how these mathematicians' conducted research and were deeply connected to their theological beliefs (Sriraman, 2004).

The mathematicians hoped that the results of their creative work would be sanctioned by a group of experts in order to get the work included in the domain (Csikzentmihalyi, 1988, 2000) primarily in the form of a publication in a prominent journal. However, the acceptance of a mathematical result, the end product of creation, does not ensure its survival in the Darwinian sense (Muir, 1988). The mathematical result, may or may not be picked up other mathematicians. If the mathematical community picks it up as a viable result, then it is likely to undergo mutations and lead to new mathematics. This however is determined by chance!

IMPLICATIONS

It is in the best interest of the field of mathematics education that we identify and nurture creative talent in the mathematics classroom. "Between the work of a student who tries to solve a difficult problem in mathematics and a work of invention (creation)...there is only a difference of degree" (Polya, 1954). Creativity as a feature of mathematical thinking is not a patent of the mathematician! (Krutetskii, 1976). Most studies on creativity have focussed on eminent individuals (Arnheim, 1962; Gardner, 1993, 1997; Gruber, 1981). The author is suggesting that contemporary models from creativity research can be adapted for studying non-eminent samples such as high school students. Such studies would reveal more to the mathematics education research community about creativity in the classroom. Educators can ask themselves (a) Does mathematical creativity manifest in the school classroom? (b) How can the teacher identify creative work? One plausible way to answer these questions is by reconstructing and evaluating student work as a unique evolving system of creativity (Gruber & Wallace,

2000) by incorporating some of the facets suggested by Gruber and Wallace (2000). This necessitates the need to find suitable problems for the appropriate levels that stimulate student creativity. A common trait among mathematicians is to rely on particular cases, isomorphic re-formulations, or analogous problems that simulate the original problem situations in their search for a solution (Polya, 1954; Skemp, 1986). It is also the case that creating original mathematics requires a very high level of motivation, persistence and reflection, all of which are considered indicators of creativity (Amabile, 1983; Policastro & Gardner, 2000; Gardner, 1993). The literature suggests that most creative individuals tend to be attracted to complexity, which most school math curricula has very little to offer. Classroom practices and math curricula rarely use problems with an underlying mathematical structure and allow students a prolonged period of engagement and independence to work on such problems. The author conjectures that in order for mathematical creativity to manifest in the school classroom, students should be given the opportunity to tackle non-routine problems with complexity and structure, which require not only motivation and persistence but also considerable reflection. This implies that educators should recognize the value of allowing students to reflect on previously solved problems and draw comparisons between various isomorphic problems (English, 1991, 1993; Hung, 2000; Maher & Kiczek, 2000; Maher & Martino, 1997; Maher & Speiser, 1996; Sriraman, 2003; Sriraman 2004b). In addition, encouraging students to look for similarities in a class of problems also fosters "mathematical" behavior (Polya, 1954), leading some students to discover fairly sophisticated mathematical structures and principles in a manner akin to creative mathematicians.

REFERENCES

Amabile, T. M. (1983). Social psychology of creativity: A componential conceptualization. *Journal of Personality and Social Psychology, 45*, 357–376.

Arnheim, R. (1962). *Picasso's Guernica.* Berkeley: University of California Press.

Birkhoff, G. (1969). Mathematics and psychology. *SIAM Review, 11*, 429–469.

Burton, L. (1984). Mathematical thinking: The struggle for meaning. *Journal for Research in Mathematics Education, 15*, 35–49.

Corbin, J., & Strauss, A. (1998). *Basics of Qualitative Research.* Thousand Oaks, CA: Sage.

Csikszentmihalyi, M. (1988). Society, culture, and person: A systems view of creativity. In R. J. Sternberg (Ed.), *The nature of creativity: Contemporary psychological perspectives* (pp. 325–339). Cambridge University Press.

Csikszentmihalyi, M. (2000). Implications of a systems perspective for the study of creativity. In R. J. Sternberg (Ed.), *Handbook of creativity* (pp. 313–338). Cambridge University Press.

Davis, P. J., & Hersh, R. (1981). *The mathematical experience*. New York: Houghton Mifflin.

English, L. D. (1991). Young children's combinatoric strategies. *Educational Studies in Mathematics, 22,* 451–474.

English, L. D. (1993). Children's strategies in solving two- and three-dimensional combinatorial problems. *Journal for Research in Mathematics Education, 24*(3), 255–273.

Ernest, P. (1991). *The Philosophy of Mathematics Education*, Briston, PA: Falmer.

Ernest, P. (1994). Conversation as a metaphor for mathematics and learning. *Proceedings of the British Society for Research into Learning Mathematics Day Conference*, Manchester Metropolitan University (pp. 58–63). Nottingham: BSRLM.

Ervynck, G. (1991). Mathematical creativity. In D. Tall (Ed.). *Advanced Mathematical Thinking* (pp. 42–53). Kluwer Academic.

Frensch, P., & Sternberg, R. (1992). *Complex Problem Solving: Principles and Mechanisms*. Mahwah, NJ: Erlbaum.

Gallian, J. A. (1994). *Contemporary Abstract Algebra*. Lexington, MA: Heath.

Gardner, H. (1993). *Frames of mind*. New York: Basic Books.

Gardner, H. (1997). *Extraordinary minds*. New York: Basic Books.

Gruber, H. E.(1981). *Darwin on man*. Chicago: University of Chicago Press.

Gruber, H. E., & Wallace, D. B. (2000). The case study method and evolving systems approach for understanding unique creative people at work. In R. J. Sternberg (Ed.), *Handbook of creativity* (pp. 93–115). Cambridge University Press.

Hadamard, J. W. (1945). *Essay on the psychology of invention in the mathematical field*. Princeton University Press.(page references are to Dover edition, New York 1954).

Hanna, G. (1991). Mathematical proof. In D. Tall (Ed.). *Advanced Mathematical Thinking* (pp. 54–60). Kluwer Academic Publishers.

Hung, D. (2000). Some insights into the generalizations of mathematical meanings. *Journal of Mathematical Behavior, 19,* 63–82.

Krutetskii, V. A.(1976). *The psychology of mathematical abilities in school children* .(J. Teller, trans. & J. Kilpatrick & I. Wirszup, Eds.). Chicago: University of Chicago Press.

L'Enseigement Mathematique. (1902), *4,* 208–211, and (1904), *6,* 376.

Lester, F. K. (1985). Methodological considerations in research on mathematical problem solving. In E. A. Silver, *Teaching and learning mathematical problem solving. Multiple research perspectives* (pp. 41–70). Hillsdale, NJ: Erlbaum.

Maher,C. A., & Kiczek R. D. (2000). Long Term Building of Mathematical Ideas Related to Proof Making. Contributions to Paolo Boero, G.Harel, C. Maher, M.Miyasaki. (organisers) *Proof and Proving in Mathematics Education. ICME9-TSG 12*. Tokyo/Makuhari, Japan.

Maher ,C. A., & Speiser M. (1997) How far can you go with block towers? Stephanie's Intellectual Development. *Journal of Mathematical Behavior 16*(2), 125–132.

Maher, C. A., & Martino A. M. (1996) The Development of the idea of mathematical proof: A 5-year case study. *Journal for Research in Mathematics Education, 27*(2), 194–214.

Manin, Y. I.(1977). *A course in mathematical logic*, New York: Springer-Verlag,

Minsky, M. (1985). *The society of mind*. New York: Simon & Schuster.

Muir, A. (1988). The psychology of mathematical creativity. *Mathematical Intelligencer, 10*(1), 33–37.

Nicolle, C. (1932). *Biologie de l'invention,* Paris: Alcan.

Patton, M. Q. (2002). *Qualitative Research and Evaluation Methods.* Thousand Oaks: Sage.

Poincaré, H. (1948). *Science and Method.* New York: Dover.

Policastro, E., & Gardner, H. (2000). From case studies to robust generalizations: An approach to the study of creativity. In R. J. Sternberg (Ed.), *Handbook of creativity* (pp. 213–225). Cambridge University Press.

Polya, G. (1945). *How to solve it.* Princeton, NJ: Princeton University Press.

Polya, G. (1954). *Mathematics and plausible reasoning: Induction and analogy in mathematics* (Vol. II). Princeton, NJ: Princeton University Press.

Schoenfeld, A. H. (1985). *Mathematical problem solving.* New York: Academic Press.

Skemp, R. (1986). *The psychology of learning mathematics.* Penguin Books.

Sriraman, B. (2003). Mathematical giftedness, problem solving, and the ability to formulate generalizations. *The Journal of Secondary Gifted Education. XIV*(3), 151–165.

Sriraman, B (2004). The influence of Platonism on mathematics research and theological beliefs. *Theology and Science, 2*(1), 131–147.

Sriraman, B. (2004). Discovering a mathematical principle: The case of Matt. *Mathematics in School, 33*(2), 25–31.

Sternberg, R. J. (1979). *Human intelligence: perspectives on its theory and measurement.* Norwood, NJ: Ablex.

Sternberg, R .J. (1985). *Human abilities: An information processing approach.* New York: W. H. Freeman.

Sternberg, R. J. (2000). *Handbook of creativity.* Cambridge University Press.

Sternberg. R. J., & Lubart, T. I. (1996). Investing in creativity. *American Psychologist, 51,* 677–688.

Sternberg. R. J., & Lubart, T. I. (2000). The concept of creativity: prospects and paradigms. In R. J. Sternberg (Ed.), *Handbook of creativity* (pp. 93–115). Cambridge University Press.

Taylor, S. J., & Bogdan, R. (1984). *Introduction to qualitative research methods: The search for meanings.* New York: Wiley.

Torrance, E. P. (1974). *Torrance tests of creative thinking: Norms-technical manual.* Lexington, MA: Ginn.

Ulam, S. (1976). *Adventures of a mathematician .* New York: Scribners.

Usiskin, Z. P. (1987). Resolving the continuing dilemmas in school geometry. In M. M. Lindquist, & A. P. Shulte (Eds.) *Learning and teaching geometry, K–12: 1987 Yearbook* (pp. 17–31). Reston, VA: National Council of Teachers of Mathematics.

Wallas, G. (1926). *The art of thought.* New York: Harcourt, Brace & Jovanovich.

Weisberg, R.W. (1993). *Creativity: Beyond the myth of genius.* New York: Freeman.

Wertheimer, M. (1945). *Productive Thinking.* New York: Harper.

Wittgenstein, L. (1978). *Remarks on the Foundations of Mathematics* (Revised Edition), Cambridge, Massachusetts Institute of Technology Press.

APPENDIX A: INTERVIEW PROTOCOL

The interview instrument was developed by modifying questions from questionnaires in L'Enseigement Mathematique (1902) and Muir (1988).

1. Describe your place of work and your role within it.
2. Are you free to choose the mathematical problems you tackle or are they determined by your work place?
3. Do you work and publish mainly as an individual or as part of a group?
4. Is supervision of Research a positive or negative factor in your work?
5. Do you structure your time for mathematics?
6. What are your favorite leisure activities apart from mathematics?
7. Do you recall any immediate family influences, teachers, colleagues or texts, of primary importance in your mathematical development?
8. In which areas were you initially self-educated? In which areas do you work now? If different, what have been the reasons for changing?
9. Do you strive to obtain a broad overview of mathematics, not of immediate relevance to your area of research?
10. Do you make a distinction between thought processes in learning and research?
11. When you are about to begin a new topic, do you prefer to assimilate what is known first or do you try your own approach?
12. Do you concentrate on one problem for a protracted period of time or on several problems at the same time?
13. Have your best ideas been the result of prolonged deliberate effort, or have they occurred when you were engaged in other unrelated tasks?
14. How do you form an intuition about the truth of a proposition?
15. Do computers play a role in your creative work (mathematical thinking)?
16. What types of mental imagery do you use when thinking about mathematical objects?

Questions regarding foundational and theological issues have been omitted in this protocol. The discussion resulting from these questions are reported in Sriraman (2004).

ACKNOWLEDGMENT

Reprint of Sriraman, B. (2004). The characteristics of mathematical creativity. *The Mathematics Educator, 14*(1), 19–34. Reprinted with permission from Mathematics Education Student Association at The University of Georgia. ©2004 Bharath Sriraman.

NOTES

1. In all vignettes I = interviewer; A,B,C,D,E = mathematicians

CHAPTER 2

MATHEMATICAL GIFTEDNESS, PROBLEM SOLVING AND THE ABILITY TO FORMULATE GENERALIZATIONS

The Problem-Solving Experiences of Four Gifted Students

Bharath Sriraman
The University of Montana

ABSTRACT

Complex mathematical tasks such as problem solving are an ideal way to provide students opportunities to develop higher order mathematical processes such as representation, abstraction, and generalization. In this study, nine freshmen in a ninth-grade accelerated algebra class were asked to solve five non-routine combinatorial problems in their journals. The problems were assigned over the course of three months at increasing level of complexity. The generality that characterized the solutions of the five problems was the pigeonhole (Dirichlet) principle. The four mathematically gifted students were

Creativity, Giftedness, and Talent Development in Mathematics, pages 33–60

33

successful in discovering and verbalizing the generality that characterized the solutions of the five problems, whereas the five non-gifted students were unable to discover the hidden generality. This validates the hypothesis that there exists a relationship between mathematical giftedness, problem-solving ability, and the ability to generalize. This paper describes the problem-solving experiences of the mathematically gifted students, how they formulate abstractions and generalizations, with implications for acceleration and the need for differentiation in the secondary mathematics classroom.

INTRODUCTION

A fascinating aspect of human thought is the ability to generalize from specific experiences and to form new, more abstract concepts. The Principles and Standards (NCTM, 2000) calls for instructional programs that emphasize problem solving with the goal of helping students develop sophistication with mathematical processes such as representation, mathematical reasoning, abstraction and generalization. It goes on to proclaim that students' should develop increased sophistication with mathematical processes, especially problem- solving, representation and reasoning, and their increased ability to reflect on and monitor their work should lead to greater abstraction and a capability for generalization. Thus, the ability to generalize is the result of certain mathematical experiences, an important component of mathematical ability and to develop this ability is an objective of mathematics teaching and learning (NCTM, 2000).

Psychologists have also been interested in the phenomenon of generalization and have attempted to link the ability to generalize to measures of intelligence (Sternberg, 1979) and to complex problem-solving abilities (Frensch & Sternberg, 1992). Greenes (1981) claimed that mathematically gifted students differed from the general group in their abilities to spontaneously formulate problems, flexibility in data management, and in their ability to abstract and generalize. There is also empirical evidence of differences in generalization in gifted and non-gifted learners at the pre-school level (Kanevsky, 1990). At the secondary level, there are very few studies that document and describe how gifted students approach problem solving, abstract and generalize mathematical concepts. This leads to the following questions:

1. What are the problem-solving behaviors that high school students engage in?
2. What are the differences in the problem-solving behaviors of gifted and non-gifted students?
3. How do gifted students abstract and generalize mathematical concepts?

DEFINITIONS

Problem-solving situation: A problem-solving situation may be defined as a situation involving:

- A conceptual task,
- the nature of which the subject by previous learning (Brownell, 1942; Kilpatrick, 1985), or by organization of the task (English, 1992), or by originality (Birkhoff, 1969; Ervynck, 1991), is able to understand,
- for which at that time, the subject knows no direct means of satisfaction.
- The subject experiences perplexity in the problem situation, but does not experience utter confusion.
- An intermediate territory in the continuum which stretches from a puzzle at one extreme to the completely familiar and understandable situation at the other (Kilpatrick, 1985).

Generalization: is the process by which one derives or induces from particular cases. It includes abstracting properties (Davis & Hersh, 1981), identifying commonalties (Dreyfus, 1991) and expanding domains of validity (Davydov, 1990; Dienes, 1961; Polya, 1954).

Problem-solving strategies: refer to the actions and/or methods employed by students in order to understand and solve the problem situation. In this study, student strategies are classified according to Lester's (1985) conceptual model on problem-solving behavior described in the literature review.

LITERATURE REVIEW

A well-known problem-solving model is that of the eminent mathematician George Polya (1945), which consists of four phases, namely, understanding, planning, implementing, and looking back. One of the shortcomings of Polya's model was that it was algorithmic in nature and research generated by it focused purely on heuristics. Lester (1985) attributed the failure of most instructional efforts to improve student's problem solving performance to an overemphasis of heuristic skills while ignoring "managerial skills necessary to regulate one's activity (metacognitive skills) " (p. 62). It was suggested that metacognitive activity or knowledge of one's thought processes or self-regulation underlay the application of heuristics and algorithms (Lester, 1985; Schoenfeld 1985, 1992). As a result, Lester (1985) modified Polya's (1945) model to include a cognitive and a metacognitive component. In the cognitive component, the four phases of understanding, planning, implementing, and looking back were relabeled as orientation, organization, execution, and verification respectively. The metacognitive component consisted of three types of variables, namely person variables, task variables and strategy variables. The four cognitive categories are described as follows.

Orientation refers to strategic behavior to assess and understand a problem. It includes comprehension strategies, analysis of information, initial and subsequent representation, and assessment of level of difficulty and chance of success. *Organization* refers to identification of goals, global planning, and local planning. The category of *execution* refers to regulation of behavior to conform to plans. It includes performance of local actions, monitoring progress and consistency of local plans, and trade-off decisions (speed vs. accuracy). Finally, *verification* consists of evaluating decisions made and evaluating the outcomes of the executed plans. It includes evaluation of actions carried out in the orientation, organization and execution categories.

The metacognitive component of Lester's (1985) model is comprised of three classes of variables, namely person variables, task variables, and strategy variables. Person variables refer to an individual's belief system and affective characteristics that may influence performance. Task variables refer to features of a task, such as the content, context, structure, syntax and process. For example, an individual's awareness of features of a task influences performance. Finally, strategy variables are those that refer to an individual's awareness of strategies that help in comprehension, organizing, executing plans, and checking and evaluation. These metacognitive behaviors are associated with the four cognitive categories. The aim of Lester's (1985) conceptual model is to describe the behaviors in the four cognitive stages in terms of "points" where metacognitive actions occur during problem solving. Philosophers sometimes describe this metacognitive action as "thinking about your thinking."

Schoenfeld (1985, 1992) suggested that problem solving must be studied within the broader context of what learning to "think mathematically" means. He described thinking mathematically as developing a mathematical point of view, valuing the processes of representation and abstraction, and having the predisposition to generalize them.

Generalization is inseparably linked to the operation of abstracting" (Davydov, 1990, p. 13). According to Davydov (1990), the process of delineating a certain quality as a common one and separating it from other qualities allows the child to convert the general quality into an independent and particular object of subsequent actions. Abstraction is a process, which occurs when the subject focuses attention on specific properties of a given object and then considers these properties in isolation from the original. This might be done to understand the essence of a certain phenomena, to later apply the same theory to other applicable cases.

Early research on generalization focused on elementary school children's abilities to generalize number concepts (Davydov,1990; Dienes,1961; Shapiro, 1965). There was also considerable interest in the process of generalization among mathematics education researchers in the former Soviet Union (Davydov, 1990; Krutetskii, 1976; Shapiro, 1965).

Shapiro (1965) wrote that among mathematically gifted students:

> the development of generalizations occurs from the first examples, at the initial stages of learning. Transfer in the general form is almost merged in time with generalization and is accomplished immediately for a whole class of problems of a single type...among less capable students generalizations ripen gradually and are manifested at later stages or they do not develop at all. (p. 95)

Krutetskii (1976) analyzed the generalization ability of both "normal" and "capable" (gifted) students in a series of experiments. He hypothesized that "students with different abilities are characterized by differences in degree of development of both the ability to generalize mathematical material and the ability to remember generalizations" (p. 84). Krutetskii (1976) studied 19 students with varying mathematical abilities. Based on his experiments with the 19 students, Krutetskii (1976) concluded that more "capable" (gifted) students were able to rapidly and broadly form mathematical generalizations. He noted that these "capable" students were able to discern the general structure of the problems before they solved them. The "average" students were not always able to perceive common elements in problems, and the "incapable" students faired poorly in this task. In order for students to correctly formulate generalizations, they had to abstract from the specific content, and single out similarities, the structures and relationships (Krutetskii, 1976).

Most of the existent literature on the process of generalization is in the context of number concepts, arithmetic and algebra. There is lack of research on generalization in the context of higher order mathematical processes such as problem solving at the high school level. In particular there is lack of research on the differences in problem-solving behaviors between gifted and non-gifted students. Such research would be valuable to practitioners' who need to differentiate the curriculum in classrooms comprising of both gifted and non-gifted students.

METHODOLOGY

The researcher in this study was a full time teacher at a rural mid-western high school. The participants in this study were nine freshmen (4 males and 5 females) enrolled in accelerated algebra I taught by the researcher. All nine students enrolled in the accelerated algebra course were willing to participate in the study. The participants were white, with middle-class socioeconomic backgrounds. Enrollment in accelerated algebra at this high school required recommendation from eighth grade teachers, as well as above average performance in pre-algebra.

The researcher did not access the testing profiles of the nine students in the class during data collection and data analysis. However, after the data was collected and analyses were constructed, the researcher accessed the testing profiles of the nine students to find that four of the nine students had been identified as mathematically gifted in their elementary school. Identification at the elementary school was based on a variety of factors such as IQ scores (over 124), the Stanford Achievement Test (95 percentile), teacher recommendations, and counselor recommendations. Table 2.1 shows the achievement profiles of the nine students.

Journal writing was an integral part of the accelerated algebra course. The teacher routinely assigned one non-routine problem or puzzle every

TABLE 2.1   Achievement Profiles of the Nine Students

Name	IQ[a]	Stanford Achievement Test Math Score[b] Raw Score (out of 90)	Stanford Achievement Test Math Applications[c] Raw Score (out of 30)	OLSAT Non-verbal[d] Raw Score (out of 36)
Subset A: Mathematically gifted students who formed generalizations				
Amy	162	89	30	36
John	124	85	29	32
Matt	140	87	30	33
Hanna	126	85	28	32
Subset B: Non-gifted students who formed false generalizations				
Bart	100	68	19	22
Jim	120	74	21	25
Isabel	105	70	20	22
Subset C: Non-Gifted students who did not form generalizations				
Jamie	98	60	15	20
Heidi	102	62	16	21

[a] Stanford-Binet (4th edition). Mean = 100 ; Standard deviation = 16; Data in columns 2–3 is extracted from the Stanford Achievement Test Series (administered to students in the 1st grade)

[b] The math portion consisted of 90 items on number concepts (34), computation (26) and applications (30)

[c] The mathematics applications portion consisted of 30 items on problem-solving (12), graphs (3) geometry (6) and measurement (9)

[d] The Otis-Lennon School Ability Test, 7th edition, administered to students in grade 6. Non-verbal portion of test consists of items on figural reasoning (18) and quantitative reasoning (18) involving mathematical concepts

other week, which the students solved in their journals. The researcher asked the students to record everything they tried, including scratch work in their journal. Full credit was given to all students for completing the three cues and including all their scratch work in their journals. Students were given three cues from the researcher:

1. Restate the problem in your own words. In other words, what is the problem asking?
2. How would you begin solving the problem?
3. Solve the problem and write a summary of what worked and what did not work.

The journal writings over the course of the school year revealed that most students were articulate in their description of their solution strategies, and were capable of tackling mathematical problems that were not covered by the school curriculum. In order to keep the setting of the study as natural and non-intrusive as possible and to be consistent with established classroom practices, the researcher assigned the five combinatorial problems (see Appendix A) for the study as journal assignments, starting with the problem of lowest complexity. These five problems were assigned over a period of three months.

The journal problems were chosen with great care and represented situations that would facilitate representation, reasoning, abstraction and eventually the formulation of generalizations. The following criteria were applied to determine the journal problems:

1. The problems had to be "mathematically" rich in the sense that they were non-routine and solving them would require both perseverance and creativity on the student's part.
2. The chosen problems were combinatorial in nature because mathematics education research with elementary school children has consistently indicated that children have intuitive abilities in tackling combinatorial problems (English, 1992).
3. The problems represented diverse situations and increased in complexity. The plan of gradually increasing the complexity of the problems was consistent with earlier research on the process of generalization (Davydov, 1990; Dienes, 1961; Krutetskii, 1976).
4. Both the problems and methods of solutions were generalizable. In addition, the solutions to a class of seemingly different problems was characterized by an over aching common generality namely the pigeonhole principle, which states that if m pigeons are put in n pigeon holes and if $m > n$, then at least one pigeonhole will have more than one pigeon.

The researcher conjectured that the strategies developed by the students would evolve with the complexity of the problem, depending on the mathematical sophistication of the students, and eventually lead some students into discovering the general principle that could be applied to all of them.

Data was collected through students' journal writings, clinical interviews and the teachers' journal writings, in the second semester of the school year. The student's were assigned the five combinatorial problems as journal problems, starting with problem 1. The rationale for providing the three cues was to initiate the four phases of problem solving (Lester, 1985). Students were given a week to ten days to solve each problem. The researcher collected the journals weekly in order to read the solutions developed by the students. The researcher then recorded in his journal possible questions to ask the students in the interview.

Students were interviewed the week after the journals were turned in before or after school. The researcher followed the clinical interview technique attributed to and pioneered by Piaget (1975) to study the thinking processes of the students. The interviews were open-ended with the purpose of getting students to verbalize their thought processes while solving a given problem. In all five rounds of interviews were conducted with the nine students over the course of three months. Students were asked questions along the following lines:

1. How did you start the problem?
2. How long did you spend on this problem?
3. How does this problem compare to the algebra problems we are doing in class right now?
4. How can you be sure that your solution is correct?
5. How would you explain your solution to a friend?
6. Did you use a known procedure to solve the problem?
7. For the second, third, fourth and fifth problems, the student was asked if he/she used a refinement/modification of an earlier strategy, and whether they could detect any similarities in the problems or their solutions?

These questions were formulated with the hope of enabling the students to verbalize their solution strategies. The researcher also wanted the students to justify their solutions and explain their reasoning. Questions 3 and 7 were explicitly asked in order to trigger generalization. After each round of interview, the researcher recorded his impressions of the interviews in his journal. The interviews were audio taped, transcribed verbatim and rechecked for errors. The data consisted of student journal writings, interview transcripts, and the researcher's journal.

The journal writing and transcribed data was coded using techniques from grounded theory (Glaser & Strauss, 1977). The constant comparative method was used in order to look for patterns in the data. The making of comparisons is an essential feature of grounded theory methodology. Four categories, namely orientation, organization, execution and verification were operationalized from Lester's (1985) problem solving model. The researcher compared action (behavior) to action (behavior) in order to classify data according to Lester's (1985) conceptual model. Each behavior was then compared to other behaviors at the property level for similarities or differences and then placed into a category. A category was characterized by properties or actions that defined or gave it meaning. When the data was coded and analyzed, similarities and differences were found in the problem-solving behaviors of the nine students as well as in behaviors that characterized the formation of generalizations for the five combinatorial problems. The categories of generalization and reflection emerged as a result of the study.

Generalization is this study was characterized as the process by which students derived or induced from particular cases. It included identifying commonalties in the structure of the problems and their solutions. It also included making analogies as well as specializing from a given set of objects to a smaller one. *Reflection* in this study was characterized as the process by which the student abstracted knowledge from actions performed on the problems. In other words reflection consisted of thinking about similarities in the problems and solutions, and abstracting these similarities over an extended time period. Finally, *affect* played an overarching role and influenced the success (failure) of the students in forming the generality that characterized the class of problems used in this study. The affective dimension includes the attitudes, beliefs, feelings, opinions, and convictions of a person (Burton, 1984; Mandler, 1984).

In order to meet concerns of validity, the researcher had triangulation of data sources, namely data from student's journal writings, interview protocols, and the researcher's journal writings. The researcher also used the strategy of intersubjectivity, by having a colleague analyze the data from the interviews by using the coding technique developed by the researcher. The colleague coded and analyzed 30 random slices of journal and interview data and came to the same conclusions as the researcher. For a given slice of data coded independently by the colleague, there was an agreement of 89% for behaviors that fell under orientation; 86% for behaviors under organization; 93% for execution; 96% for generalization; and 91% for reflection. This lends validity to the findings of this research.

The researcher met reliability concerns by studying students in the same ninth grade algebra class. The researcher documented his observations of the students over the school year in his journal. In terms of putting sufficient time in the field, since the researcher was also the teacher of the ninth grade students, he was fully immersed in the culture of the classroom. In addition, the personal interviews conducted with the students were tape recorded, transcribed verbatim and given to the students with a request for clarifications, omissions, or additions.

LIMITATIONS

The reader should be aware that the context of the study contributed to the nature of the results. Hence, the researcher would like to point out the unique characteristics of this qualitative study, so that the reader can "judge" the applicability of the results in other settings.

The students in this study were freshman in an accelerated (honors) algebra class in a rural high school. Demographically speaking they were all white, with middle class socioeconomic backgrounds. Eight out of the nine students aspired to finish high school with AP calculus. The students in the study were motivated to succeed in school and were willing participants in the research effort. Thus student motivation and willingness to participate in the study may have influenced their effort levels and the results obtained.

The nine students that participated in this study had a year of pre-algebra in the eighth grade. In their mathematical background prior to high school, these students had not been exposed to constructing mathematical proofs, nor had they been expected to construct general solutions to pre-algebraic and algebraic problems. It is conceivable that if students had been exposed to problems involving mathematical proof, they would have been able to distinguish between problems that asked for existence solutions (Problems 1 and 2) as opposed to those that asked for general solutions (Problems 3, 4, and 5).

The researcher had extremely high expectations of the students in the accelerated algebra class. He expected the students to invest considerable time independently on the journal problems. The complexity of the five problems used in this study indicates that the researcher expected extraordinary leaps of thought from high school freshmen. The researcher was also instrumental in encouraging students to write out their strategies and reflective summaries of journal problems over the course of the first semester. Thus students already had this background when the five combinatorial problems were given in the second semester. Thus the articulateness found in journal writings is attributable to researcher's influence on the students.

The students in the accelerated algebra class were told to work on the problems independently without referring to books, friends, or other people in the class. The diversity in the journal solutions, the variety of explanations during the interviews, and the inability of many students to find solutions to the last three problems indicate that the students did not collaborate with one another. However there is always a miniscule possibility that some students may have talked to each other about the problems.

RESULTS

Qualitative analysis of the nine case studies resulted in three subsets based on the problem-solving behaviors and generalizations developed by the students. Subset A included Amy, John, Matt, and Hanna, who were successful in discovering the generality that characterized the solutions of the problems, namely the pigeonhole principle. They were able to isolate the similarities in the structure of the problems and their solutions. They discerned that the solutions entailed two unequal quantities be "matched up" or compared, and then were able to point out how this played a role in the solution of the problem, i.e., any given slot was forced to have "more than one" of a quantity. These students showed great perseverance curiosity, and were motivated to pursue the problems and reflect on them over an extended period of time. As stated earlier the researcher accessed the testing profiles of the nine students after data collection and analysis to find that the four students in subset A had been identified as mathematically gifted in their elementary school.

Subset B included Bart, Jim, and Isabel whose overall generalization scheme was to employ algebraic operations on the given numbers in the problems. These students focused on the superficial similarities in the problems, and tried to apply procedures from algebra. Their comparisons of the problems often showed several inconsistencies. They were unable to pursue a train of thought from one interview to the next over the course of the five problems.

Finally, subset C included Jamie and Heidi, whose overall generalization scheme was "finding numerous examples that work" and in some problems this was modified to include "numerous examples that don't work." This general scheme was used over and over again. These students had the attitude of wanting to get the problem over with, and were primarily concerned with executing and verifying the scenarios in the problem situations.

The similarities and differences in the behaviors of the students in the three subsets are presented in Tables 2.2–2.4. This is followed by a descriptive section, which focuses on some of the solutions of Amy, John, Matt and Hanna, and includes interview vignettes that reveal the problem-solving strategies and the mathematical experiences of these four gifted students. Table 2.2 compares the problem-solving behaviors of the students

TABLE 2.2 Comparisons of Student Problem-Solving Behaviors in the Orientation, Organization, Execution, and Verification Phases

	Subset A Gifted (Amy, John, Matt, Hanna)	Subset B Non-Gifted (Bart, Jim, Isabel)	Subset C Non-Gifted (Jamie, Heidi)
Orientation	Consistently comprehending the problem situation.	Miscomprehending the problem situation.	Miscomprehending the problem situation.
	Assessing the adequacy of information given in the problem situation.	Listing the "given" numbers in a problem situation.	
	Identification of (and understanding) the assumptions of the problem situation.	Making up assumptions for the given problem situation.	Poor understanding of assumptions for the given problem situation.
	Distinguishing between interrogative and declarative statements.	Unclear distinction between interrogative and declarative statements.	No distinction between interrogative and declarative statements.
Organization	Global planning. Consistently planning to work their "way up" or "starting out small."	Haphazard/vague global planning.	Local planning.
Execution	Controlling the variability of the problem situation.	Not controlling the variability of the problem situation	Not controlling the variability of the problem situation
	Performance of correct local actions.	Performance of "unusual" local actions.	Performance of local actions.
	Continuously monitoring progress and consistency of plans.	Not carefully monitoring progress and consistency of plans.	Monitoring progress and consistency of plans.
Verification	Checking results of local actions.	Inconsistencies in results of local actions.	Use of one particular case for verification.
	Verifying consistency of results with implemented plans.	Inconsistency of results with implemented plans.	Use of examples/non-examples to reach conclusions.
	Use of particular cases to better understand why a phenomenon occurred.	Use of particular cases to verify if a phenomenon occurred.	

TABLE 2.3 Comparisons of Student Behaviors in Generalization and Reflection

	Subset A Gifted (Amy, John, Matt, Hanna)	Subset B Non-Gifted (Bart, Jim, Isabel)	Subset C Non-Gifted (Jamie, Heidi)
Generalization	Identifying similarities in the structure of the problems.	Identifying superficial similarities in the structure of the problems.	Identifying superficial similarities in the structure of the problems.
	Identifying similarities in the solutions of the problems.	Inconsistencies in verbalization of similarities in the solutions of the problems.	Contriving similarities in the solutions of the problems.
	Using analogical reasoning.		
	Refining methods where appropriate.	Forcing connections with concepts in algebra.	
	Extending the domain of validity.		
	Verbalizing common principles.	Articulation barriers.	
Reflection	Conjecturing and examining plausible examples and non-examples.	Conjecturing but not examining the plausibility of a conjecture.	Little or no conjecturing.
	Relating to previous experience.	Poor decision making during execution and verification.	
	Decision making during and after execution and verification.	Putting aside problem after completion.	Putting aside problem after completion.
	Thinking about similarities in the problems and solutions.		
	Abstracting structural similarities in the problems and solutions over an extended time period.	Abstracting superficial similarities from the wording of the problems and solutions.	No abstraction.

TABLE 2.4 Comparisons of Affective Behaviors

	Subset A Gifted (Amy, John, Matt, Hanna)	Subset B Non-Gifted (Bart, Jim, Isabel)	Subset C Non-Gifted (Jamie, Heidi)
Affect	Perseverance	Lack of perseverance	Lack of perseverance
	Confidence/Lack of confidence	Confidence/Lack of confidence	Lack of confidence
	Curiosity	Low degree of curiosity	Low degree of curiosity
	Excitement		
	Frustration	Satisfaction	Satisfaction
	Valuing communication	Lack of eagerness to communicate	No eagerness to communicate
	Mathematics as a "way of thinking"	Mathematics as operations on numbers	Pre-conceived notion of "problem-solving" from middle school textbook

in the three subsets in the orientation, organization, execution, and reflection phases of problem solving. Table 2.3 compares the generalization and reflective behaviors of the students in the three subsets. Finally Table 2.4 compares the affective behaviors of the students in the three subsets. The purpose of these tables is to allow the reader to compare the problem solving behaviors of the gifted students (subset A) with those of the non-gifted students (subsets B and C).

The Mathematical Experiences of the Gifted Students

Problem #1: The Soda Problem

Hanna began the problem by restating it in her own words. She wrote, "There are 6 listed sodas in Blaise's Bistro. If one student placed an order for 1 soda, how many students would have to place an order so that one of the six sodas would be ordered by at least two students? In other words, how many students would it take to order a soda per student so that one soda was ordered at least twice?"

Hanna's plan to solve the problem was "by making a list of the six different sodas. She would then write "1st student, 2nd student etc to symbolize 1 student per order . . . and so on for all six sodas." Finally Hanna's solution consisted of a list with the six sodas. She assigned one student per soda for the six listed sodas, and then assigned the "7th student across from ginger ale to symbolize the second student for one of the 6 sodas." From her work, Hanna "gathered that 7 students are required to order a soda, one soda per student, to insure that at least one of the six sodas would get ordered

Days	Pills	Days	Pills	Days	Pills
1	1	11	1	21	2
2	1	12	1	22	2
3	1	13	1	23	2
4	1	14	1	24	2
5	1	15	2	25	2
6	1	16	2	26	2
7	1	17	2	27	2
8	1	18	2	28	2
9	1	19	2	29	2
10	1	20	2	30	1

Figure 2.1 Matt's representation of the aspirin problem.

by at least two students." The researcher was impressed by Hanna's journal. She had restated the problem in her own words, made a plan, executed the plan, and then stopped once she had verified that her solution had fulfilled all the problem conditions.

Problem #2: The Aspirin Problem

Matt began the problem by writing, "The problem is asking to find a sequence of aspirin taken in 30 days and find out if in a number of consecutive days he will take 14 aspirin." He understood this to mean "there must be a way for the person to take 14 aspirin in any amount of consecutive days" and to make sure that all of the 45 aspirins were gone in 30 days. In order to solve the problem Matt's strategy was to make a list of the 30 days, and then to write the number of aspirin on the other side (see Figure 2.1).

Matt executed his plan by making a chart with 30 days on one side and the number of aspirin on the other. "I will try to write at least one pill in one day and put two pills on some days to make the 45 pills. I believe this should work." Matt assigned one pill for each of the 30 days and then assigned an additional pill for days 15 thorough 29. He then wrote, "It is possible, the least amount of days it takes to get 14 aspirin is 7 days." There was no further reflection on Matt's part, and no acknowledgement of the fact that other solutions were possible. He also seemed to be convinced that the answer was 7 days based on his solution. However, the following vignette reveals that Matt was aware of other solutions to the problem in addition to being able to identify the structural similarities in the first two problems.

Vignette 1

(In all vignettes S = Student; I = Researcher/Teacher)

> **I:** Do you think this is the only way to do it?
> **S:** I think so.

I: So there is no other way?

S: You mean, are these like the only kind of numbers you can use?

I: Yeah.

S: No, cause you can use a couple of 3's. Or you can put one day with a whole lot of them, and the rest of the days you can just have ones.

I: Okay, how long did you spend on the problem?

S: I looked at it for a couple of days before, and I thought about it for a while, and then in my study hall I wrote the first two tasks. Then I went home, and the next day I spent about half an hour trying to figure out how to do it.

I: So you think you spent more time on this compared to the first one?

S: Yeah, the first one was a bit too easy.

I: Do you see any similarities in the two problems?

S: Mmm . . . What I noticed was going through and putting one, then adding one for some many days and you get the answer.

I: Okay

S: That's what I did for all of them (pointing to the soda problem), one had to have two, so I put two for one of them.

I: And with the second problem?

S: I put in one in all of them, and then I changed half the ones to twos.

Problem #3: The Number Sum Problem

Amy recorded her thoughts on what the problem was asking in her journal by writing:

> This problem is asking *how* something happens. I have to figure out how this happens and prove to you and myself that this will always happen. This may seem impossible, but it will always work . . . it's very interesting . . . you would think that it wouldn't work in some set of numbers, but it works in each and every set.

In order to start the problem Amy plan was to make up a set of ten numbers between 1 and 100. She would then this set to find two ways of getting the same sum. She would then make up another set and keep repeating this process. She wrote, "Hopefully I will see a pattern. Then I can prove how this happens."

Amy's first set was {3, 12, 23, 29, 53, 61, 70, 79, 81, 94}, and the two sums that she very quickly found was: 3 + 12 + 79 = 94, a member in the set, in other words the second sum is simply 94 = 94. Interestingly enough, her method for finding this sum was to first subtract 79 from 94 to get 15, and then to note the fact that the sum of 3 and 12 was 15. Hence 3 +12 + 79 = 94.

Her second set was {9, 12, 29, 41, 45, 71, 73, 88, 97, 98}, and this time her sum was 45 + 12 + 41 = 98. At this point, she decided to try something different. She would start with a set of ten numbers, but now revise it somehow, i.e., to change some elements, with the hope of getting "different results." She wrote that at this stage she was just experimenting with the hope of discovering something. She began with the set {5, 14, 16, 29, 44, 46, 53, 61, 80, 89} and looked for different ways to get a solution, in other words, sums that equaled an element in the set, or two sums that were equal. The different solutions that she found were 5 + 14 + 61 = 80; 46 + 5 + 29 = 80. She decided then to "substitute" some numbers and "find new results." So she changed the numbers 5, 14, 29, 46 and 80 to different numbers. She started with the set {5, 14, 16, 29, 44, 46, 53, 61, 80, 89, which, she labeled "original" set, and changed it to {6, 7, 16, 21, 44, 49, 53, 61, 82, 89}, which she labeled the "revised" set. She substituted 6, 7, 21, 49, and 82 for the numbers 5, 14, 29, 46 and 80, in order to see if this would result in a set, which did not give any solutions. However, she quickly found more sums and expressed surprise that it still worked. " 7 + 82 = 89; Wow, I found one right away!" She decided to find more sums and found a number of sums, such as 7 + 21 + 61 = 89, a member of the revised set; 16 + 49 = 21 + 44; 61 + 21 = 82; and 7 + 16 + 21 = 44. Having reached another dead end, she decided to try something new. She wrote, " I am going to put some thought into choosing my 10 numbers."

She finally created the set {1, 2, 4, 8, 16, 32, 64,...} as follows: "I started out with 1, then I chose 2. Now, I didn't want to get a solution, so my next number obviously wasn't going to be 3, so I put 4 instead, because 1 + 2 = 3 and I would have had a solution already. I continued working this way. The next bigger number would be 7, because 1 + 2 + 4 = 7, so I didn't choose 7, instead I chose the next number 8." She continued this pattern of carefully choosing the numbers, until she came to 64. She also observed that she was doubling the previous number to choose the next number. "There is still no possible solutions but if I doubled 64, I would get 128 . . . and the problem states that the numbers have to be from 1–100, so I couldn't use 128. I still have 3 numbers to go. Now, if I pick any random number, look what happens." Amy picked numerous random numbers between 64 and 100, such as 87, 99, 68, 71, 84, 92 and always found sums that equaled these numbers. For example: 87 = 64 + 16 + 2 + 1 + 4. She then picked numbers between the numbers in her carefully chosen set, such as 5, 17, 45, 50, 9 and 29, and she found sums that equaled these numbers. Her conjecture was that 7 numbers were the maximum that one could have in a set without finding a solution, namely two sums that were equal. She came to the conclusion that seven integers were the most one could choose without finding solutions, and since the problem asked to choose ten integers, there would always be a solution.

She had abstracted the fact that going beyond a maximal number resulted in a certain phenomena, which fulfilled the conditions of the problem. In the first problem, going beyond 6 students forced one soda to be ordered twice. In the second problem, going beyond 30 pills, forced the person to take more than one pill a day, and resulted in a string of consecutive days where exactly 14 pills were consumed. Now, in the case of the ten number set, Amy's creation of a maximal set with 7 carefully chosen elements, resulted in two sums that were equal when an eighth element was chosen. It was at this point that Amy reflected on her solution, and wrote, "Since I tried to get the most variation as possible in the set {1, 2, 4, 8, 16, 32, 64,...}, it reminds me of the first problem, where I tried to get the most variety of orders possible, assigning one person to each soda. *So after all, these two problems are similar!* Did you plan this?"

It was clear that Amy was beginning to develop an intuitive grasp of the hidden generality in the problems, namely the pigeon hole principle. During the interview, she was able to identify the structural similarities in the three problems and verbalize the pigeonhole principle.

Problem #4: The Acquaintance Problem

John understood this problem to be asking "I could take twenty people and prove whether or not the same person would have the same amount of friends as somebody else." In order to solve the problem John wrote that he would label the twenty people and then use "guess and check." To solve the problem John drew the following (Figure 2.2):

1	2	3	4	5	6	7	8	9	10
1	2	3	4	5	6	7	8	9	10

11	12	13	14	15	16	17	18	19	20
11	12	13	14	15	16	17	18	19	1–19

Figure 2.2 John's representations of the acquaintance problem.

He explained his figure as follows. The top row stands for the people in the room, i.e., 1 stands for person 1 and so on. The second row stands for the number of friends. For example, according to the table, person 1 had one friend, person 2 had two friends etc. John wrote, " I thought for a while and decided if 1 knew 1 friend, then 2 knew 2 friends and so on. I tried not to have the same friend amount twice."

The researcher conjectured that John actually reflected on the problem before deciding to proceed along the lines of not assigning the same number of friends to the people in the room. Proceeding along these lines when he came to person 20 in the room, he deduced that person since person 20 couldn't possibly know 20 people in the room, he or she would have to know at least one person, or at most 19 people. Hence two people would end up with the same number of friends.

The researcher found this argument remarkable, because it applied the pigeonhole principle implicitly, in order to conclude that two people had to end up with the same number of friends. The researcher asked John to explain his solution strategy and then asked him if there were any similarities in the problems or the solutions to the problems, which eventually led to verbalization of the pigeonhole principle, the generality, which characterized his solutions to Problems #1, 2 and 4. The following vignette illustrates John's discovery of the pigeonhole principle.

Vignette 2

S: I made a table for all of them, and I listed all the people and brands possible, and figured it out by placing numbers to where it works.

I: So how is the solution the same?

S: A table and list the people and figure out the numbers for each specific person.

I: Is that the only thing in common?

S: Among the first, second and fourth. Yeah. It's all asking for a specific amount . . . like for people or brand of soda, Like they ask for a certain amount that can be possible. For this one (pointing to the aspirin problem) the certain amount could be twenty.

I: You think there is some kind of an equation, a rule that you talked about before, that is common to these?

S: There are 6 soda pops and 7 people, 30 days and 45 aspirin. I am breaking down the numbers.

I: So what is going on?

S: Like there are more people than soda pops and more aspirin than days, and there are more friends known than people.

I: More friends known then people?

S: Oh! Yeah . . . one is shorter.

I: What is shorter?

S: More people than friends. They are never equal. There is always one thing that is more.

I: So how does all that help you solve the problem?

> **S:** The problems say at least one, so I knew there would be
> more people than sodas.
> **I:** And what happened as a result?
> **S:** (Silence). (Writing down numbers) . . . The problem
> could then be solved.

At this point in the interview John had correctly identified the similarities in the structure of the first, second and fourth problem. He verbalized this by saying that all the problems were asking for a certain amount, or if a certain amount was possible? He realized that he had made a table for his solutions of 1, 2, and 4, and that there were two quantities being compared, with the characteristic that one quantity was "shorter" than the other. He also said that this allowed the problems to be solved. Since John had implicitly stated what enabled the problems to be solved, namely the pigeonhole principle, the researcher made the pedagogical decision of using John's numbers to facilitate the verbalization of the pigeonhole principle. It is important that the reader note that the researcher made this decision only after the student had implicitly stated the pigeonhole principle in their own words. In John's case this was evident after he had identified the two quantities that were being compared, with one being shorter than the other, which enabled the problem to be solved. This was his way of stating the pigeonhole principle. This will become evident to the reader in the following vignette.

Vignette 3

> **I:** Since you threw all those numbers at me 6 and 7, 30 and
> 45, 19 and 20 . . . why don't I ask you this. What if you used
> these numbers for holes and these for pigeons?
> **S:** (Laughing) There will always be one more pigeon?
> **I:** Where?
> **S:** There will be two in a hole?
> **I:** So tell me what happens here?
> **S:** Then it is going to be . . . you could put 15 in one and put
> the rest in the others (pointing to the numbers 30 and
> 45). Some holes are going to have more than one pigeon?
> **I:** What about these numbers (pointing to the 19 and 20)
> **S:** One hole will have two here.
> **I:** You still didn't tell me that equation of yours?
> **S:** (A long period of silence). Pigeons greater than holes.
> Let x be the holes, then the pigeons will be greater than
> the number of holes.
> **I:** How could you say that using your x?
> **S:** $1 + x$

I: So you wrote down that you have x holes and $1 + x$ pigeons and that would?

S: Then all the holes are filled, and one more than once . . . I was surprised how it worked for the last problem (the acquaintance problem) . . . that worked out.

Thus, John arrived at the generality that characterized his solutions of Problems #1, 2, and 4. During this journey the researcher noted that John had reflected on a possible "equation" for three weeks. Unlike Amy who just stumbled upon the generality during the course of solving the third problem, John consciously looked for something that would capture the problems in an equation. Note that John was very quick in his verbalization of the pigeonhole principle which is attributable to his implicit use of this principle in his solutions to three of the four problems.

Problem #5: The Band Problem

None of the nine students were able to solve the band problem. Amy and Matt constructed plausible explanations about how the shuffles occurred but were unable to apply the pigeonhole principle to solve this problem. The solutions to the five problems are located in Appendix B.

This concludes the descriptive narratives of the problem-solving experiences of the four gifted students. Amy stumbled on the pigeonhole principle after solving the first three problems and was able to apply it to solve Problem #4. John verbalized the pigeonhole principle after attempting the first four problems, by reflecting on and identifying the similarities in Problems #1, 2, and 4. Finally Matt and Hanna discovered the pigeonhole principle after attempting all five problems, by identifying the similarities in Problems #1, 2, and 4. Among the gifted subset Amy was by far the most successful student. Her creation of a maximal set in trying to solve the third problem was also deemed mathematically noteworthy by several math professors at a nearby university.

CONCLUSIONS AND IMPLICATIONS

The results indicate that the four gifted students, Amy, John, Matt and Hanna were successful in forming generalizations. Broadly speaking the major difference between the gifted students and the others lay in the orientation, organization and reflection phases of problem solving. The gifted students invested a considerable amount of time in trying to understand the problem situation, identifying the assumptions clearly, and devising a plan that was global in nature. Although these students never found general solutions to the five problems, they did consistently work their "way up" by beginning with simpler cases that modeled the given problem situation. In doing so, they were controlling the variability of the problem situation. In other words they realized that the quantities in the given problems were not invariant. For example: In Problem #3, Amy controlled the variability of the problem by picking the integers in their sets more and more carefully. In doing so they did not restrict themselves to a set with just ten integers but tried sets with less than ten integers. During the execution phase of problem solving these students consistently performed correct local actions (manipulations) and monitored their progress. Once the results of local actions were obtained, they were checked for accuracy and consistency. In the verification phase of last three problems, these students consistently made use of particular cases to gain an insight into why a phenomenon occurred. In terms of forming generalizations for this class of problems, the successful students correctly identified similarities in the structure of three or more problems as well as similarities in their solutions. They were adept at using analogies during their explanation of the similarities in the problems. They were also able to communicate effectively and verbalize the common principle that they thought characterized three or more of the problems. In many instances they specialized from consideration of a given set of objets to a smaller sub-set, contained in the given one.

The generalization behaviors exhibited by Amy, John, Matt and Hanna show several consistencies with the existing research literature. Krutetskii (1976) came to the conclusion that in order for students to correctly formulate generalizations, they had to abstract from the specific content, and single out similarities, the structures and relationships. Amy, John, Matt and Hanna were able to do this in varying degrees. The reflective behaviors of these students consisted of thinking about similarities in the problems and solutions, and abstracting these similarities over an extended time period. This finding also verifies the conjectures of Piaget (1971) and Dubinsky (1991), who viewed generalization as a process of "reflective abstraction." Dubinsky (1991) claimed that generalization involved the combination of objects and processes, which involved a high degree of awareness of the part of the subject. In this study the five problems were the objects and the solutions to the five problems were the processes, and the generality that characterized the class of problems and solutions was the pigeonhole principle.

There were other reflective behaviors that were shown by the gifted students, which are not explicitly mentioned in the research literature as aiding the process of generalization. In the context of this research study, there was a considerable amount of decision-making behavior exhibited by the gifted students during and after the execution and verification phases of problem solving. Decision-making could be viewed as "quick" reflective behavior during the problem solving process that steered the students towards a correct solution. Another reflective behavior was conjecturing after attempting a given problem. After the students had attempted a given problem they often formed a conjecture, and then pursued their conjecture by examining plausible examples. This was a striking feature of Amy's journal writings and also observed in the others. For instance John said that he was looking for an equation that would solve the problems, and he finally came up with it after solving Problem #4.

The researcher mentioned earlier that there were subtle differences in the quality of generalizations formed by the gifted students. Amy was the most successful student in this subset because she discovered the generality after Problem #3, and was also able to solve Problem #4 by using the generality she had constructed. She then tried to subsume Problem #5 under the general method but was unable to do so. In a way she was operating at the level of a mathematician. A mathematician would view Amy's generalization scheme as follows. First a general method was derived from Problems #1, 2 and 3. The method was then formulated explicitly (as the pigeonhole principle) and considered as an entity by itself, and its structure was analyzed. This structure was then used to further include a different type of problem (Problem #4), without making changes to the original method (Skemp, 1986). John and Matt on the other hand were able to arrive at the generality after solving Problem #4 and Problem #5 respectively, but were unable to subsume Problems #3 and 5 under the generality. In Hanna's case she arrived at an intuitive understanding of the generality by abstracting similarities in Problems #1, 2 and 4, and by consciously excluding Problems #3 and 5 from this process. Since abstraction is a premise of generalization (Davis & Hersh, 1981; Dayvdov, 1990) the abstracting behaviors of these students are similar to those exhibited by mathematicians and enabled to be successful in forming valid generalizations.

The affective behaviors of the gifted students in forming generalizations are consistent with the literature (Burton, 1984; Mandler, 1984). According to Burton (1984) cognitive activity is charted by affective responses that can be observed as passing through three phases: *entry, attack, and review*. The phase of engaging a problem is called entry. Surprise, curiosity or tension creates an affective need that is resolved by exploration (attack), which in turn satisfies the cognitive need to get a sense of the underlying pattern, which in this study was the pigeonhole principle. In the gifted students the researcher noted the powerful positive emotions that often went along with the construction of new ideas (Glasersfeld, 1987). This result is consistent with research literature claims that most affective factors arise out of emotional responses to the interruption of plans or planned behavior (Burton, 1984; Mandler, 1984; Schoenfeld, 1985). Besides showing surprise and curiosity, over the course of the five problems the gifted students also showed remarkable perseverance and experienced bouts of frustration. They valued communication and viewed mathematics as a "way of thinking."

The four gifted students had a natural predisposition (Shapiro, 1965) to engage in the manifested problem-solving and generalization behaviors. Despite not being offered enrichment and acceleration opportunities during the middle school years, it was interesting to find that the gifted students showed a high level of reflective behavior in addition to concern with orientation and organization in a problem-solving situation. This finding should be of great interest to teachers and counselors. The gifted students displayed conceptual understanding because they were able to abstract similarities and form valid conceptual links. Affect played a major role in how they approached a problem situation. In particular their beliefs about what constituted mathematics influenced how they tackled a given problem. The gifted students may have found the problems captivating enough that it created an affective need for them to get a sense of the underlying pattern that characterized them (Burton, 1984; Mandler, 1984; Schoenfeld,1985).

So, if teachers want students to become adept at forming generalizations, the first and foremost challenge is to find various classes of problems which have general solutions, that are accessible to the students and which capture their interest. The findings indicate that gifted students are particularly capable of abstracting similarities in the structure of problems and situations in a manner akin to mathematicians, as well as in formulating valid mathematical generalizations. This makes it crucial for high school teachers to create learning opportunities that allow mathematically gifted students to develop and apply their talents.

REFERENCES

Birkhoff, G. (1969). Mathematics and psychology. SIAM Review, 11, 429–469.

Brownell, W. A. (1942). The place and meaning in the teaching of arithmetic. *The Elementary School Journal, 4,* 256–265.

Burton, L. (1984). Mathematical thinking: The struggle for meaning. *Journal for Research in Mathematics Education, 15,* 35–49.

Davis, P. J., & Hersh, R. (1981). *The mathematical experience.* New York: Houghton Mifflin.

Davydov, V. V. (1990). *Type of generalization in instruction: Soviet studies in mathematics education.* Reston, VA: National Council of Teachers of Mathematics.

Dienes, Z. P. (1961). On abstraction and generalization. *Harvard Educational Review,* 31, 281–301.

Dreyfus,T. (1991). Advanced mathematical thinking processes. In D. Tall (Ed.), *Advanced mathematical thinking* (pp. 25–40). The Netherlands: Kluwer Academic Publishers.

Dubinsky, E. (1991). Constructive aspects of reflective abstraction in advanced mathematics. In L. P. Steffe (Ed.) *Epistemological foundations of mathematical experience* (pp. 160–187). New York: Springer-Verlag.

English, L. D. (1992). Problem solving with combinations. *Arithmetic Teacher,* 40(2), 72–77.

Ervynck, G. (1991). Mathematical creativity. In D. Tall (Ed.). *Advanced mathematical thinking* (pp. 42–53). The Netherlands: Kluwer Academic Publishers.

Frensch, P. & Sternberg, R. (1992). *Complex problem solving: principles and mechanisms.* Mahwah, NJ: Erlbaum.

Gardner, M. (1997). *The last recreations.* New York: Springer-Verlag.

Glaser, B & Strauss, A. (1977). *The discovery of grounded theory: Strategies for qualitative research.* San Francisco, CA: University of California San Francisco.

Glasersfeld, E.von. (1987). Learning as a constructive activity. In C. Janvier (Ed.), *Problems of representation in the teaching and learning of mathematics* (pp. 3–18), Hillsdale, NJ: Lawrence Erlbaum Associates.

Greenes, C. (1981). Identifying the gifted student in mathematics. *Arithmetic Teacher,* 28, 14–18.

Kanevsky, L. S. (1990). Pursuing qualitative differences in the flexible use of a problem solving strategy by young children. *Journal for the Education of the Gifted,* 13, 115–140.

Kilpatrick, J. (1985). A retrospective account of the past twenty-five years of research on teaching mathematical problem solving. In E. A. Silver (Ed.) *Teaching and learning mathematical problem solving. Multiple research perspectives* (pp. 1–16). Hillsdale, NJ: Lawrence Erlbaum and Associates.

Krutetskii, V. A. (1976). *The psychology of mathematical abilities in school children.*(J. Teller, trans. And J. Kilpatrick & I. Wirszup, Eds.). Chicago: University of Chicago Press.

Lester, F. K. (1985). Methodological considerations in research on mathematical problem solving. In E. A. Silver (Ed.), *Teaching and learning mathematical problem solving. Multiple research perspectives* (pp. 41–70). Hillsdale, NJ: Lawrence Erlbaum and Associates.

Mandler, G. (1984). *Mind and body: psychology of emotion and stress.* New York: Norton.

National Council of Teachers of Mathematics. (2000). *Principles and standards for school mathematics.* Reston, VA: Author.

Piaget, J. (1971). *Biology and knowledge.* Edinburgh University Press.

Piaget, J. (1975). *The child's conception of the world.* Totowa, NJ: Littlefield, Adams.

Polya, G. (1945). *How to solve it.* Princeton, NJ: Princeton University Press.

Schoenfeld, A.H. (1985). *Mathematical problem solving.* New York: Academic Press.

Schoenfeld, A.H. (1992). Learning to think mathematically: problem solving, metacognition, and sense making in mathematics. In D.A. Grouws (Ed.). *Handbook of research on mathematics teaching and learning* (pp. 334–368). New York: Simon and Simon Schuster.

Shapiro, S.I. (1965). A study of pupil's individual characteristics in processing mathematical information. *Voprosy Psikhologii, 2,* 1–113.

Skemp, R. (1986). *The psychology of learning mathematics.* Penguin Books.

Sternberg, R. J. (1979). *Human intelligence: perspectives on its theory and measurement.* Norwood, NJ: Ablex.

ACKNOWLEDGMENT

Reprint of Sriraman, B. (2003). Mathematical giftedness, problem solving, and the ability to formulate generalizations. *The Journal of Secondary Gifted Education,* 14(3), 151–165. Reprinted with permission from Prufrock Press. ©2003 Bharath Sriraman

APPENDIX A—THE PROBLEMS

Problem 1

The soda menu of a Bistro has six choices for sodas, namely Cola, Diet Cola, Lemonade, Ginger Ale, Root Beer, and Diet Root Beer.

How many students would be required to place soda orders, one soda per student, in order to insure that at least one of the six listed sodas would be ordered by at least two students?

Problem 2

A person takes at least one aspirin a day for 30 days. Suppose he takes 45 aspirin altogether. Is it possible that in some sequence of consecutive days he takes exactly 14 aspirin? Justify your solution.

Problem 3 (adapted from Gardner, 1997)

Choose a set S of *ten positive integers* smaller than 100.

For example I choose the set $S = \{3, 9, 14, 21, 26, 35, 42, 59, 63, 76\}$

There are two completely different selections from S that have the same sum.

For example, in my set S, I can first select 14, 63, and then select 35, 42, Notice that they both add up to 77. $(14 + 63 = 77; 35 + 42 = 77)$

I could also first select 3,9, 14 and then select 26. Notice that they both add up to 26. $(3 + 9 + 14 = 26;$ and $26 = 26)$

No matter how you choose a set of *ten positive integers* smaller than 100, there will always be two completely different selections that have the same sum.

Why does this happen? Prove that this will always happen.

Problem 4

There are 20 people in the room. Some of them are acquainted with each other, some not. Prove that there are two persons in the room who have an equal number of acquaintances.

Problem 5 (adapted from Gardner, 1997)

A rectangular array consists of rows and columns. Consider a marching band whose members are lined up in a rectangular array of m rows and n columns (m and n can be any natural numbers). Viewing the band from the left side, the bandmaster notices that some of the shorter members are hidden in the array. He rectifies this aesthetic flaw by arranging the musicians in each row in increasing order of height from left to right so that each one is of height greater than or equal to that of the person to his left (from the viewpoint of the bandmaster). When the bandmaster goes around to the front, however he finds that once again some of the shorter members are concealed. He proceeds to shuffle the musicians within their columns so that they are arranged in increasing order of height from front to back.

At this point he hesitates to go back to the left side to see what his adjustment has done to his carefully arranged rows. When he does go, however, he is pleasantly surprised to find that the rows are still arranged in increasing order of height from left to right! Shuffling an array within its columns in this manner does not undo the increasing order in its rows. Why does this happen? Prove that this will always be the case.

APPENDIX B—THE SOLUTIONS

Problem 1

The Soda problem has the obvious solution that seven students would be required to place soda orders, since the worst case scenario is each of the first six students order a different drink, thus forcing the seventh student to order a drink that has been previously ordered.

Problem 2

The Aspirin problem is commonly resolved by assuming that the person takes at least one aspirin pill a day. Therefore in thirty days, the person would have consumed exactly thirty aspirin pills, thus leaving a surplus of fifteen pills, which the person can randomly take in the thirty-day cycle. Let a_i be the total number of aspirin consumed up to and including the ith day, for $i = 1, \ldots, 30$. Combine these with the numbers $a_1 + 14, \ldots, a_{30} + 14$, providing 60 numbers, all positive and less or equal to $45 + 14 = 59$. Hence two of these 60 numbers are identical. Since all a_is and, hence, $(a_i + 14)$s are distinct (at least one aspirin a day consumed), then $a_j = a_i + 14$, for some $i < j$. Thus, on days $i + 1$ to j, the person consumes exactly 14 aspirin.

Problem 3

The Number Sum problem can be solved as follows. There are $2^{10} = 1,024$ subsets of the 10 integers, but there can be only 901 possible sums, the number of integers between the minimum and maximum sums. With more subsets than possible sums, there must exist at least one sum that corresponds to at least two subsets. Hence there are always two completely different selections that yield the same sum.

Problem 4

The Acquaintance problem can be resolved as follows. If there is a person in the room who has no acquaintances at all then each of the other persons in the room may have either 1, or 2, or 3,... or 18 acquaintances, or do not have acquaintances at all. Therefore we have 19 "holes" numbered 0, 1, 2, 3,..., 19, and have to distribute between them 20 people. Next, assume that every person in the room has an acquaintance. Again, we have 19 holes 1, 2, 3,..., 19 and 20 people. Thus two people will be forced to have the same number of acquaintances.

Problem 5

The Band problem can be proved via *reductio ad absurdum* (proof by contradiction). Assume that all the columns have been arranged but there is a row in which a taller musician A (column I) has been placed in front (or to the left) of a shorter musician B (column J). Since the columns have been arranged every musician in segment X from A back in column I is at least as tall as A, and every musician in segment Y from B forward in column J is no taller than B. Since A is taller than B, this implies that members in segment X are taller than members in segment Y. Now consider the halfway point at which the rows have been arranged but not the columns. To get to this point one must move the musicians from segment X to their former positions throughout column I and return to segment Y to their positions through column J. The members of X and Y have to be distributed over the rows 1, 2,..., m, as if the m rows were the holes. The segments X and Y have total length of $m + 1$. By the pigeonhole principle two musicians must end up in the same row. They could not come from the same segment, so in some row there must be a member C from X, ahead of a member D from Y. Since C is taller than D, this arrangement violates the established increasing order of rows. So the conclusion follows by *reductio ad absurdum*.

CHAPTER 3

GIFTED NINTH GRADERS' NOTIONS OF PROOF

Investigating Parallels in Approaches of Mathematically Gifted Students and Professional Mathematicians

Bharath Sriraman
The University of Montana

ABSTRACT

High school students normally encounter the study and use of formal proof in the context of Euclidean geometry. Professional mathematicians typically use an informal trial and error approach to a problem, guided by intuition to arrive at the truth of an idea. Formal proof is pursued only after mathematicians are intuitively convinced about the truth of an idea. Is the use of intuition to arrive at the plausibility of a mathematical truth unique to the professional mathematician? How do mathematically gifted students form the truth of an idea? In this study, four mathematically gifted freshmen with no prior exposure to proof nor high school geometry were given the task of establishing the truth or falsity of a non-routine geometry problem, sometimes referred to as "circumscribing a triangle" problem. This problem asks whether it is true

Creativity, Giftedness, and Talent Development in Mathematics, pages 61–84

61

that for every triangle there is a circle that passes through each of the vertices. This paper describes and interprets the processes used by the mathematically gifted students to establish truth and compares these processes to those used by professional mathematicians. All four students were able to think flexibly as evidenced in their ability to reverse the direction of a mental process and arrive at the correct conclusion. This paper further validates the use of Kruteskiian constructs of flexibility and reversibility of mental processes in gifted education as characteristics of the mathematically gifted student.

INTRODUCTION

The Principles and Standards for School Mathematics, published by the National Council of Teachers of Mathematics (NCTM) envisions classrooms in which students "make, refine, and explore conjectures on the basis of evidence and use a variety of reasoning and proof techniques to confirm or disprove those conjectures"(NCTM, 2000, p. 3). The NCTM (2000) envisions students in all grade levels approaching mathematics in a manner akin to professional mathematicians. For instance, in the early elementary level teachers are encouraged to create learning experiences that allow students to develop pattern-recognition and classification skills, to encourage students to justify their answers via use of empirical evidence and short chains of deductive reasoning grounded in previously accepted facts. As students progress to middle grades they are expected to have frequent experiences with formulating generalizations and conjectures, evaluating conjectures and constructing mathematical arguments. Finally in high school students are expected to become adept at working formally with definitions, axioms, theorems and be able to write proofs.

The recommendations outlined by the *Principles and Standards* are generic and meant to apply to all students. However there is a substantial body of research in gifted education, which indicates that mathematically gifted students are different from their peer groups in many ways. For instance mathematically gifted students differ from their peers in their capacity for learning at a faster pace (Chang, 1985; Heid, 1983), and in their curiosity to understand the conceptual ideas (Johnson, 1983; Sheffield, 1999). Further they differ from their peers in their ability to abstract and generalize (Greenes, 1981; Kanevsky, 1990; Krutetskii, 1976; Shapiro, 1965; Sriraman, 2002, 2003a), in information processing abilities, data management (Greenes, 1981; Yakimanskaya, 1970), in flexibility and reversibility of operations (Krutetskii, 1976) and in their tenacity and decision making abilities in problem solving situations (Frensch & Sternberg, 1992; Sriraman, 2003a). Instructional studies at Stanford University (Suppes & Binford, 1965; Goldberg & Suppes, 1972) showed that with instruction talented students can master inference principles such as modus ponens, modus tollens, and hypothetical syllogism, which are all pre-cursors to proof, as early as the 5th grade!

This led the author to hypothesize that mathematically gifted students may have an intuitive notion of proof and its role in mathematics even if they had never had any prior instruction on proof. In other words do mathematically gifted students have a natural capacity to approach proof in a manner akin to mathematicians? Usually mathematicians first form a personal belief about the truth of an idea and use this as a guide for more formal analytic methods of establishing truth. For example, a mathematician may intuitively arrive at the result of a theorem, but realize that deduction is needed to establish truth publicly (Fischbein, 1980; Kline, 1976; Polya, 1954). Thus, intuition convinces the mathematician about the truth of an idea while serving to organize the direction of more formal methods (Fischbein, 1980), namely the construction of a proof to "publicly" establish the validity of the finding (Bell, 1976; Manin, 1977; Mason et al., 1992).

This leads to the following questions:

1. How do mathematically gifted students arrive at an intuition of truth?
2. How do mathematically gifted students convince themselves and others about their intuition of truth ?
3. Do the approaches used by the gifted students parallel those used by professional mathematicians? If so, what are the parallels?

LITERATURE REVIEW

Epp (1990) states that the kind of thinking done by mathematicians in their own work is "distinctly different from the elegant deductive reasoning found in mathematics texts." (p. 257). This statement puts into perspective the challenges faced by a student when expected to construct deductive arguments upon first encountering geometry in high school. When one talks to a mathematician about mathematical discovery, mathematicians acknowledge making illogical steps in arguments, wandering around in circles (Lampert, 1990), trying guesses (Davis & Hersh, 1981; Poincaré, 1948) and looking at analogous examples (Fawcett, 1938; Polya, 1954) for help, and yet the end result does not give the student this insight into the hidden struggle beneath the crisp, dry proof.

Chazan (1993) examined high school geometry students' justification for their views of empirical evidence and mathematical proof and reported his findings from in depth interviews with seventeen high school students from geometry classes that employed empirical evidence. The focus of Chazan's (1993) analysis was on students' reasons for viewing empirical evidence as proof and mathematical proof simply as evidence. In the first part of the interview students were asked to compare and contrast arguments based on the measurement of examples and deductive proof. The second part of the interview focussed on the textbook deductive proof and sought to clarify if interviewees believed that a deductive proof proves the conclusion true for all objects satisfying the given. They were also asked to draw counterexamples if they could. Chazan (1993) concluded that students had a good reason to believe that *evidence is proof* in the realm of triangles because there was sufficient evidence to support the claim. These students expressed skepticism of the ability of a deductive proof to guarantee that no counter examples exist.

The van Hiele model (1986) of geometric thought emerged from the doctoral works of Dina van Hiele-Geldof and Pierre van Hiele in the Netherlands. The model consists of five levels of understanding, which can be labeled as visualization, induction, induction with informal deduction, formal deduction and finally proof. These labels describe the characteristics of the thinking at each stage. The first level is characterized by student's recognizing figures by their global appearance or seeing geometric figures as a visual whole. Students at the second level (analysis) are able to list properties of geometric figures; the properties of the geometric figures become a vehicle for identification and description. The third level students begin to relate and integrate the properties into necessary and sufficient sets for geometric shapes. In the fourth level students develop sequences of statements to deduce one statement from another. Formal deductive proof appears for the first time at this level. In the fifth level students are able to analyze and compare different deductive systems. The van Hiele levels of geometric thinking are sequential and discrete rather than continuous, and the structure of geometric knowledge is unique for each level and a function of age. Van Hiele believed that instruction plays the biggest role in students moving from one level of geometric thinking to the next higher level. He also claimed that without instruction, students may remain indefinitely at a given level. The author does not agree with van Hiele's claim that the levels are discrete and a function of age. This claim may hold true for non-gifted students but they certainly do not apply to mathematically gifted students as will be argued in the next paragraph. Morever the van Hiele model does not take a holistic view of mathematical ability and is strictly confined to the realm of geometry.

There were numerous experiments conducted in the former Soviet Union in the time period 1950–1970 (Ivanitsyna, 1970; Krutetskii, 1976; Menchinskaya, 1959; Shapiro, 1965; Yakimanskaya, 1970) with mathematically "capable" students which demonstrated that gifted students have a repertoire of abilities which cannot be pigeonholed into discrete levels within a narrow sub-domain of mathematics such as Euclidean geometry..

Instead these researchers characterized the mathematical abilities of gifted children holistically as comprising of analytic, geometric and harmonic components and argued that gifted children usually have a preference for one component over the others. The analytic "type" has a mathematically abstract cast of mind, the geometric "type" has a mathematically pictorial cast of mind, whereas a harmonic type is a combination of analytic and geometric types. For instance, given the same problem, one gifted child might pursue an analytic approach, whereas another would pursue a geometric approach. Strunz (1962) gave a different classification of 'styles' of mathematical giftedness and suggested the "empirical type" and the "conceptual type." In this classification the empirical type would have a preference for applied situations, immediately observable relations and induction, whereas the conceptual type would have a preference for theoretical situations and deduction. Krutetskii (1976) observed that one of the attributes of mathematically gifted students was the ability to switch from a direct to a reverse train of thought (reversibility), which gifted students performed with relative ease. The mathematical context in which this reversibility was observed was in transitions from usual proof to proof via contradiction (reductio ad absurdum), or when moving from a theorem to its converse. The researchers cited in this paragraph acknowledged the use of 'intuitive' ability in gifted children. In the author's knowledge there are no studies that have looked at how gifted students use their intuition in mathematics. There are however a limited number of studies with mathematicians which have tried to increase our understanding or how mathematicians use intuition (Fischbein, 1980; Kline, 1976; Sriraman, 2004).

Kline (1976) found that a group of mathematicians said they began with an informal trial and error approach guided by intuition. It is this process which helped these mathematicians convince themselves about the truth of a mathematical idea. After the initial conviction formal methods were pursued:

> The logical approach to any branch of mathematics is usually a sophisticated, artificial reconstruction of discoveries that are refashioned many times and then forced into a deductive system. The proofs are no longer natural or guided by intuition. Hence one does not really understand them through logical presentation. (p. 451)

Fischbein (1980) believed that intuition is an essential component of all levels of argument. Fischbein (1980) referred to the use of intuition as anticipatory. "While trying to solve a problem one suddenly has the feeling that one has grasped the solution even before one can offer an explicit, complete justification for that solution" (Fischbein, 1980, p. 10).

Sriraman (2004) interviewed five mathematicians to determine qualitative characteristics of creative behavior. In this study the mathematicians were questioned about how they formed an intuition of the truth of a proposition. All of the mathematicians in this study mentioned that the last thing they looked at was a formal proof. The mathematicians went about forming an intuition about truth by consciously trying to construct examples and counter-examples (Sriraman, 2004). In other words they worked a problem both ways, constructing examples to verify truth, as well as looking for counterexamples that would establish its falsity, thus using a back and forth approach of conscious guessing (Bell, 1976, Lampert, 1990; Polya, 1954, Usiskin, 1987). The "ideal mathematician," a fictional entity constructed by Davis and Hersh (1981), when asked by a student of philosophy "what is a mathematical proof?" replies with lots of examples such as "the fundamental theorem of this, the fundamental theorem of that. Etc." When probed by the philosophy student the ideal mathematician finally succumbs and confides that "formal logic is rarely employed in proving theorems, that the real truth of the matter is that proof is just a convincing argument, as judged by competent judges" (Hersh, 1993, p. 389). This leads us once again to the questions posed earlier about mathematically gifted students.

How do mathematically gifted students arrive at an intuition of truth? How do mathematically gifted students convince themselves and others about their intuition of truth? And do the approaches used by the gifted students parallel those used by professional mathematicians?

METHODOLOGY

The Participants

Since one of the goals of this study was to determine whether mathematically gifted students have intuitive notions about proof, it was important to select students that had no prior instruction on proof. The participants selected for this study were four freshmen in a large rural mid-western high school, enrolled in various sections of Integrated Mathematics I (a NSF funded curriculum aligned to NCTM Standards, developed at Western Michigan University). The author was a full time mathematics teacher and gifted coordinator of this rural mid-western high school. The four students had been previously enrolled in the same K–8 school district, one of the feeder schools of the high school, and had been identified as mathematically gifted by the K–8 district based on test scores on the Stanford Achievement Test (95 percentile), and teacher nominations. The author was provided with this information by the K–8 district. Table 3.1 gives the achievement profiles of the four students and shows that they were in the top one percentile on the Stanford Achievement Test.

TABLE 3.1   Testing Profiles of the Four Students

	Stanford Achievement Test (Grade 1) Math Raw Score[a] (out of 90 items)	Stanford Achievement Test (Grade 8) Math Raw Score[b] (out of 82 items)	Percentile Rank (National)
Jill	90	82	99
Yuri	89	80	99
Kevin	89	81	99
Sarah	88	80	99

[a] The math portion of the 8th edition consisted of 90 items subdivided into items that measured number concepts (34), computation (26) and applications (30).
[b] The math portion of the 9th edition consisted of 82 items subdivided into items that measured problem solving (52) and procedures (30).

In addition, at the end of their first semester in the high school, their respective ninth grade math teachers identified and recommended the four students for the high school gifted program. The achievement profiles in mathematics along with the nomination by math teachers at the K–8 district as well as the high school ascertain the mathematical giftedness of these four students.

The four students were not enrolled in any of the math courses taught by the author. They were invited to participate in this study through a letter stating that the gifted coordinator (author) was interested in studying the mathematical thinking of gifted students. The four students consented to participate in this study and completed a survey about their previous K–8 coursework in mathematics and answered specific questions about their familiarity with geometry and proof. The surveys indicated that their previous math course was Algebra with some enrichment. All four students indicated studying classification of geometric figures based on properties in grades 4 and 6. One of the students mentioned an interest in geometric constructions but reported no instruction was given in school. The mathematics curricula in grades 7 and 8 did not include any instruction on Euclidean geometry or proof. The only content that related to geometry and proof respectively were a small unit on the use of formulas to determine surface areas and volumes of geometric shapes, and an enrichment unit on establishing identities in ratio and proportion.

The Problem

The problem selected for this experiment is sometimes referred to as the "circumscribing a triangle problem." A perusal of commonly used textbooks in high schools indicated that this problem is normally encountered as an enrichment problem towards the end of the school year in geometry. This problem is also found in analytic geometry books because it can be solved using analytic and/or algebraic tools.

The problem states:
Consider the triangle below. The circle passes through each vertex of this triangle.

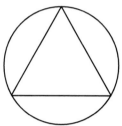

  a. Is it true that for every triangle there is a circle that passes through each of the vertices?
  b. If yes (why?). If no, how would you go about finding out?

This problem was deemed suitable for an extended investigation because of the following reasons.

  1. The problem was very simply stated, easy to understand, and the four students had not encountered such a problem before. Therefore they were confronted with a novel task.
  2. The problem presented visual information on the basis of which false inferences could be made.
  3. The problem could be approached in a variety of ways, algebraically, analytically, empirically, via geometric construction, and logically, thereby allowing for various styles of solutions to manifest.
  4. The problem was generally stated although a particular case was presented in the figure.

Data Collection Procedures

The clinical interview technique attributed to and pioneered by Piaget (1975) to study the thinking processes of the students was followed. Each student was individually interviewed after school. The interviews were task based, centered on the aforementioned problem. The interviews were open-ended with the purpose of getting students to verbalize their thought processes while solving the given problem. In all four interviews were conducted, each lasting approximately an hour. The students were probed at length by the author and asked to "think aloud." The following questions were asked:

1. How would you go about convincing someone who thinks that the statement is (the opposite of what the student has said).
2. How would one find the center and/or radius of the circle circumscribing a triangle?
3. If the student based inferences on the given figure, they were asked why?
4. What constituted a proof in mathematics?

The students were asked to explain their reasoning in great detail. The interviews were audio taped, transcribed verbatim and re-checked for errors. The students were provided with a copy of the interview transcript and asked to make clarifications they thought were necessary. The objective of doing this was not to misconstrue what the students said, to have a complete and accurate interview transcript, and to ensure compatibility between what students said and what they had meant to say. The four students were satisfied with the clarity of the transcripts and did not make any clarifications or corrections on the transcripts. In addition, the author recorded his impressions about the interview immediately after each interview. The data consisted of interview artifacts (student work), interview transcripts, and the author's notes.

Data Coding and Data Analysis

The transcribed data was coded and analyzed using techniques from grounded theory (Glaser & Strauss, 1977). Coding began by reading the interview transcripts line by line and spontaneously memoing words that described the mental processes employed by the four students. The goal of coding was to delineate the processes and build categories (Strauss & Corbin, 1998). The author purposefully looked for actions that corresponded to a process, noting its evolution as students responded to the problem. The constant comparative method was applied to compare the actions of the four students and to isolate the similarities of their thought processes as found in the data. The following categories emerged as a result of data coding and analysis.

The category of visualization emerged when students repeatedly verbalized the visual information provided indicating that the circumscribed triangle was equilateral. There were 108 memos of words and phrases such as "it looks equilateral," "the angles and sides look equal," "it looks like a perfect triangle," etc. The category of intuition emerged as a result of 137 memos of words like "It just seems right... I don't know why?" "it just seems obvious," "I'm sure there is a way...," etc. In other words these memos pointed to assertions of self-evidence. There were 212 memos for words indicating measurement and use of concrete examples, which led to the creation of the category of empiricism. Finally there were 82 memos for phrases hinting at a reversal of the process of looking at the problem. Phrases such as "how can I fit points inside...," "What if I started with the circle...," etc. which led to the category of reversibility. The four categories are now defined.

Definitions

- *Visualization:* The process by which a student makes inferences by transforming or inspecting pictures. (Hershkowitz, 1989).
- *Empiricism:* is referred to the repeated use of examples that provide conforming (non-conforming) evidence in order to support the truth of an idea. Also involves use of specific measurements to make inferences (Chazan, 1993; Polya, 1954, Szmitz, 1982).
- *Intuition:* The affective mood associated with having grasped the solution while trying to solve a problem "before one can offer an explicit and complete justification for that solution." (Fischbein, 1980; Kline, 1976). Involves the use of reasoning that is not formal, use of everyday terms and invoking empirical examples for purposes of justification (Poincaré, 1948; Polya, 1954).
- *Reversibility:* The process (or ability) to switch from a direct chain of thought to a reverse chain. The ability to reverse a mental operation (Krutetskii, 1976). This includes the ability to solve (or think about) the same problem in several different ways.

Validity

The author used the strategy of intersubjectivity (Rubin & Babbie, 1997), by having a colleague analyze the data from the interviews using the coding technique developed. The colleague coded and analyzed 36 random slices of interview data and came to the same conclusions as the author. For the slices of data coded independently by the colleague, there was an agreement of 93% for processes indicating visualization, 91% for empiricism, 92% for intuition, and 96% for reversibility. This lends validity to the findings of the research effort.

RESULTS

The results of the study are first presented under the categories that emerged as a result of data coding and analysis. The categories that emerged as processes used to construct a "proof" were visualization, intuition, empiricism and reversibility. The author presents the student's pathways to "proof" in the form of tables summarizing the patterns in each category. This is followed by an extensive interpretation and commentary on the observed patterns and their isomorphisms to mathematical techniques used by professional mathematicians. Finally the author establishes validity of the findings by using "triangulation by theory"—application of various explanations from the literature to the data at hand and the selection of the most plausible ones to explain the research results.

All four students came to the conclusion that the statement was true for every triangle by an inductive process of trial and error. The process of proving the statement began with the intuition that the statement was true only for equilateral triangles (based on the visual information). The students then ascertained this truth for equilateral triangles by intuitively constructing the center and formulating counterexamples to validate their conjecture that the statement was false in general. They finally determined the truth of the statement by dramatically reversing their thinking. An overall picture of this process is found in Figure 3.1.

VISUALIZATION

Visualization played an important role in the process of establishing the truth of the statement. All four students insisted that the given triangle was equilateral because it "looked" like one. Although the statement clearly asked if it was possible to circumscribe a circle over every triangle, the students couldn't ignore the visual image. This led them to conjecture that the given statement applied to only equilateral triangles or "special" triangles. Table 3.2 provides examples of student verbalizations of their conjectures based on the visual information.

Intuition

The four students followed their initial intuition that the statement applied only to equilateral triangles. Three out of the four were also able to determine the correct construction to locate the center of the circle circumscribing an equilateral triangle. This was remarkable because they had

Intuitive construction to determine the center of the circle
circumscribing the equilateral triangle

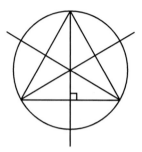

Pathologies invented by students to violate the given statement

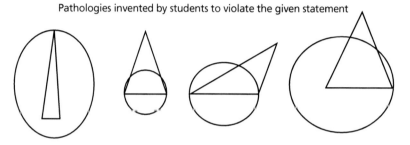

Reversing the mental process (starting with a circle first)

 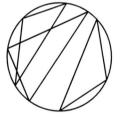

Figure 3.1 Samples of interview artifacts.

never been taught this construction before. However their intuition guided
them into discovering the construction. It is important to note that students
did not have a compass and straightedge available and all constructions
were done freehand (see Figure 3.1). Table 3.3 provides glimpses of student intuition used to construct the center of the circle.

TABLE 3.2   Seeing Is Believing

Category	Examples of process	Student
Visualization	It would only work for equilateral triangles (*pointing to the figure that looks like an equilateral triangle*).	Jill
	Right now you have an equilateral triangle. At least it looks like one.	Yuri
	It's working over here because it is an equilateral triangle.	Kevin
	This looks like an equilateral triangle with equal distances.	Sarah

TABLE 3.3   The Center of the Circle and the Equilateral Triangle Coincide...I'm Certain

Category	Examples of process	Student
Intuition	Draw the perpendiculars that pass through midpoints and then when you have the center, you can take the distance to one of the vertices as the radius and join them...*It just seems right. I don't know why?*	Jill
	I draw an altitude to each side of the triangle...Where they intersect...*it just seems obvious that this will give the center.*	Yuri
	I know there is some way to do it. *It's obvious that there is some way to find it.*	Kevin
	I'll draw the equilateral triangle, the perpendicular...and another perpendicular and the point where they cross would be the center.	Sarah

Empiricism

Students were probed as to whether the construction they had discovered applied only to equilateral triangles? This led them to construct counterexamples (see Figure 3.1) to substantiate their intuition that the statement applied only to equilateral triangles, and was false in general. Table 3.4 provides glimpses of this empirical process of constructing counterexamples used by the four students.

Reversibility

At this stage of the interview each of the four students were almost convinced that the statement was false in general. It is noteworthy that they

TABLE 3.4 Look At All These Weird Triangles

Category	Examples of process	Student
Empiricism	If you have a triangle say like this (*draws a thin scalene triangle*). You couldn't get a circle to pass through this triangle. It would be more like an ellipse.	Jill
	You can't always use the altitudes to determine the center. Let me draw another triangle.	Yuri
	Now if you take a really different type of triangle then it won't work. Here is a triangle and it is not working.	Kevin
	Yeah, I tried fitting a circle around these other triangles, it didn't work.	Sarah

TABLE 3.5 Let's Start with the Circle First!

Category	Examples of process	Student
Reversibility	*Wait a minute... I suppose it is true. You can draw a circle and always fit some triangle inside it.* (*Draws an example*) You can fit it as long as it is inside the circle.	Jill
	I've found a new construction, what is stopping me from fixing the base points elsewhere? *I can fix the two points elsewhere (drawing chords) and then choose the third point. Yes, the statement is true.*	Yuri
	Let me try something else. I'll draw a really queer triangle and I'm going to make it look like this (draws an obtuse scalene triangle). Would this work? *But if you pick these 3 points on the circle, it seems to work... Yeah!* You can always pick the points on the circle and then draw the triangle.	Kevin
	I'm trying to think here (tearing the paper in frustration). *What if I follow the circle and pick points?* (Pause) Yeah I'm trying to visually look at the triangles and I... guess the fact that no matter what kind of triangle I draw, *if I can draw a circle first then I can draw any triangle in it* (trying more examples). *I must draw the circle first. Yes, it's true, it's a true statement.*	Sarah

weren't willing to commit to saying that the statement was false in spite of the counterexamples they had constructed. The students wanted to try a different approach, which was evidence of their flexibility in thinking, a trait of mathematically gifted students (Kruteskii, 1976). Table 3.5 shows similarities in how students dramatically reversed their thinking by starting with an arbitrary circle first instead of the triangle. By reversing their train of thought they were able to convince themselves that the statement was indeed true.

INTERPRETATIONS AND ISOMORPHISMS

In the preceding tables and figure, similarities in the student pathways to their "proof" were re-constructed. The author contends that the thinking processes of the four students show remarkable isomorphisms to those of professional mathematicians as will be discussed in this section. Mathematics is often viewed as an activity of creating relationships, some of which are based on visual imagery (Casey, 1978; Presmeg, 1986). The image presented to the four students immediately kindled the formation of a preliminary conjecture, that the given triangle was equilateral. This in turn led to the question of determining the center of the circumscribed circle which resulted in the students discovering the relationship that the center of the circle coincided with the point of intersection of the three perpendicular bisectors. It is important to understand that it is impossible to directly find out how children (and hence how mathematicians) create images, however the manner in which they use images can be studied from their actions on a given problem (Inhelder & Piaget, 1971). We as adults have often seen children do some strange things when faced with a mathematical task. One often comes across reactions from teachers where they view the child's intuitive actions as bizarre (Kamii & DeClark, 1985). This kind of reaction reflects more on the teacher's inability to imagine things from the child's perspective. Some one reading an abstract proof is in a similar situation because of being unable to imagine the proof from the creative mathematician's perspective, as well as being unaware of the images used by the mathematician in its creation. The four students were able to isolate attributes (equal sides and equal angles) that they deemed critical (Hershkowitz, 1989) in order to form the initial conjecture about the truth of the given statement. In other words the figure served as a visual reference point to kindle the mathematical process of proving.

Intuition is the guide that mathematicians use to convince themselves about the validity of a proposition (Burton, 1999; Fischbein, 1980; Kline, 1976; Sriraman,2004). For the four students, the process of proving began with the intuition that the statement was true for equilateral triangles. The author interprets this as the intuitive action of specializing the given statement for equilateral triangles. Mathematical thinking is often characterized by four processes, which are *specializing, conjecturing, generalizing*, and *convincing* (Burton, 1984; Burton, 1999). Once the students had specialized and conjectured that the statement was true for equilateral triangles, they were asked if this implied that the statement was true for all triangles. This led to a quasi-empirical (Ernest, 1991; Lerman, 1983) approach to proof, in which the students tried to construct mathematical pathologies (Figure 3.1) or mathematical monsters (Lakatos, 1976), in the form of triangles that would disobey the given proposition. This quasi-empirical process again shows remarkable similarities to a view of mathematical thinking introduced by the eminent mathematical philosopher Imre Lakatos, in which mathematics is presented as a model of possibilities, subject to conjecture, proof and refutation. In other words mathematics is not viewed as an absolute immutable body of knowledge, but is instead subjected to the scientific process of constantly revising and refining preliminary hypothesis. In this view of mathematics, no theorem or proof is perfect because there is always the possibility of a better revision. The quasi-empirical process of constructing pathologies by the four students is a common trait among mathematicians when working on problems. Pathologies serve the purpose of revisiting the problem and refining the hypothesis or assumptions.

The quasi-empirical process employed by the four students led to the revised conjecture that the statement was perhaps false in general. However at this juncture they were still unwilling to commit to saying that the statement was false in spite of the counterexamples they had constructed. A common trait among professional mathematicians is to work on a problem for a prolonged period of time, and if no break through occurs, mathematicians often stand back and "sleep on it." In other words they let the problem incubate and hope that an insight or breakthrough will eventually occur (Wallas, 1926; Wertheimer, 1945). This is the Gestalt view of mathematical thinking (Hadamard, 1945; Poincaré, 1948; Sriraman, 2004; Wallas, 1926; Wertheimer, 1945). Mathematicians often characterize this as the stage where the "problem talks to you." The author contends that this occurred in a microcosmic way with the four students. After having spent close to an hour on the problem, they put their pencils aside and mulled on the problem in silence for a few minutes. Remarkably enough the insight to reverse their thinking dawned upon them (Table 3.5). This process has many interpretations. Insightful mathematical thinking and creativity can be viewed as a process of non-algorithmic decision-making (Ervynck, 1991; Poincaré, 1948). Decisions that have to be made by mathematicians may be of a widely divergent nature and always involve a crucial choice (Birkhoff, 1969; Poincaré, 1948 ; Ervynck, 1991). It is interesting that mathematicians view a crucial aspect of their craft as non-algorithmic decision making, in an age where the use of the computing power of machines to gain insights into results is slowly becoming a valid approach. This was the most intense and frustrating stage for the four students where conceptual activity (Ervynck, 1991) occurred, and manifested in an illumination (Wallas, 1926; Wertheimer, 1945), viz., the decision or choice to reverse the structure of the problem.

It is common among mathematicians to work on the problem one day and then on its converse the next day, or simultaneously work the problem both ways to gain an insight. This process of reversibility is viewed as an aspect of flexibility in thinking, a trait of mathematically gifted students

(Kruteskii, 1976), and connects well with the back and forth approach employed by mathematicians when tackling a problem.

After the students had convinced themselves that the statement was true for all triangles, they were asked how they would convince others about this truth. In other words students were probed about the methods they would employ to establish the truth publicly. It is remarkable that all four students intuitively knew that one counterexample was sufficient to establish falsity, however establishing truth involved more work and would require "substantial" evidence. The students relied on the use of empirical evidence to explain truth and were convinced that numerous visual examples were enough to convince others about the truth of the statement. In other words, a proof for them was explaining and convincing (Bell, 1976; Kline, 1976). This is a very natural view of proof even among professional mathematicians. "The formal logic view of proof is a fascinating topic of study for logic…but it is not a truthful picture of real-life mathematical proof." (Hersh, 1993, p. 391). The views expressed by the four students about the role of proof are sophisticated for 9th graders and again show remarkable isomorphisms to that held by professional mathematicians and some philosophers of mathematics. The author will use some quotes to illustrate this to the reader:

> I look for examples that will support it and those that won't support it. I have to be looking for examples that would disprove the statement otherwise I would be wasting a lot of time doing work for nothing.…Proof is written explanation or examples explaining, based on previous things that I believe to be true. (Yuri)

This quote of Yuri is astoundingly isomorphic to the view of proof expressed by one professional mathematician, a brilliant analyst:

> First I get an idea that something along a certain line should be true and then I start to prove it, and in that proof I run into some difficulties and then I say, can I construct an example from those difficulties? If in constructing the example I run into difficulties…then can those difficulties be put into this proof you know, so I do this back and forth, usually at some gut level I have the belief that something along that line is true. Not always is that intuition correct but it is correct often enough and…I am able to prove something that I suspect is true. (Quote from a professional mathematician in Sriraman, 2004)

This next quote from Kevin shows startling similarities to Lakatos' (1976) thesis of mathematics being an ever-evolving process of conjecture proof and refutation:

> You can always find one case where it doesn't work. So to prove something is true, even like in science you can have a theory that it will work but you can

never definitely be sure that it will always work. . . . You never prove something is true. You can take a bunch of different types of cases and see if it works and then it is generally accepted that its true . . . unless someone comes around and proves it's false. There are things that are believed to be true for 200 years and someone will come around with a particular case where its' false. (Kevin)

Informal, quasi-empirical mathematics does not grow through a monotonous increase in the number of indubitably established theorems but through the incessant improvement of guesses by speculation and criticism, by the logic of proofs and refutations. (Imre Lakatos (1976), *from Proofs and Refutations*)

Finally, the use of visual arguments, by the students, to convince others about the truth of a statement has historically found expression in Indian mathematics. The following quotes illustrate this isomorphism:

I guess you would start out like this visual, get your arguments down and then put it into words. I remember a lot of times starting out visually and just seeing and working with that and being able to put it into a proof.[1] . . . (Laughing) I just don't see the point sometimes. Like, if I know its true, why do I have to go through 16 steps in proving it? It's much more effective in getting the point across with visual examples. (Sarah)

In general the mathematics of Europe was influenced by Greek mathematics while Indian mathematics, despite influences from Greece and Arabia establish a unique tradition . . . there was no conflict between, on the one hand, visual demonstration and numerical calculation and, on the other, proof by deduction. (Almeida, 2003)

Our inherited notion of rigorous proof is not carved in marble. People will modify that notion, will allow machine computation, numerical evidence, probabilistic algorithms, if they find it advantageous to do so. Then, we are misleading our pupils, if in the classroom we treat rigorous proof as shibboleth. (quoted by Almeida, 2003; original quotation in Hersh, 1993, p. 395)

TRIANGULATION BY THEORY AND IMPLICATIONS

In this study four mathematically gifted ninth grade students with no formal exposure to proof or Euclidean geometry were given the task of establishing the truth or falsity of a statement. The strategies used by the students to construct a "proof" were documented, coded and analyzed. It was found that these students relied on visualization (Hershkowitz, 1989; Yakimanskaya, 1970), empiricism involving the use of examples and counterexamples or conscious guessing (Bell, 1976; Lampert, 1990; Polya, 1954; Sriraman,

2004), and reversibility (Kruteskii, 1976; Shapiro, 1965) in order to arrive at the truth. This entire process was guided by their strong intuition as evidenced in their ability to formulate conjectures and devise constructions to validate their initial conjecture about equilateral triangles. It is noteworthy that although the four students were faced with nonconforming evidence in the form of "weird" triangles which seemed uncircumscribable, they were unwilling to commit to stating the statement was false. It is easier to say that something is false based on a poorly constructed counterexample as is evidenced in high school geometry (Senk, 1985; Usiskin, 1987), whereas to state something is true in mathematics involves the conviction that the statement holds for potentially infinitely many cases. The four gifted students were aware of this distinction whereas most high school students in geometry think otherwise and believe statements to be true just for a particular figure (Mason, 1996; Senk, 1985).

Mathematicians have an awareness of the generality of a statement by distinguishing between *looking through* and *looking at*. *Looking through* is analogous to generalizing through the particular whereas *looking at* is analogous to specializing (identifying) a particular case in the general (Dubinsky, 1991; Mason, 1996). A simplistic example that comes to mind and also given by Mason (1996) is in a high school geometry setting when a teacher draws a (particular) triangle on the board and says that the sum of the angles of a triangle is 180 degrees. Most often what is stressed is the empirical fact, namely 180 degrees. "The generality of the statement is hidden in the indefinite article *a*. A student *looking through* this statement sees the general in the particular and recognizes that the essence of the statement is the *invariance* of the angle-sum in the domain of all possible triangles. *Looking through* entails recognizing the attribute of invariance in an implied domain of generality" (Mason, 1996, p. 65). The four gifted students in this study were able to look through the statement posed in the problem and recognize the attribute of invariance, a quality of professional mathematicians. The gifted students were also aware of the differences between convincing themselves and convincing others. This was clear when they said that convincing the class entailed organizing the convincing evidence and constructing an argument in a coherent way (Hersh, 1993, Hoyles, 1997, Mason et al., 1992). They demonstrated flexibility in thinking by about the problem differently. This manifested in the remarkable way they reversed their strategy (Krutetskii, 1976) to conclude that the statement was indeed true.

In terms of Strunz's (1962) classification of styles of mathematical giftedness, the four gifted students showed a preference for immediately observable relations and induction but were conceptually aware that proving a statement entailed all possible cases. They could reason in the abstract and would be classified as amalgams of the "empirical" and "conceptual" types.

If one utilized the holistic classification of the Soviet researchers, the four students' mathematical giftedness was of the harmonic type, a combination of the analytic and geometric types. They were able to use their pictorial representations to analytically induce the truth of the statement (Ivanit-syna, 1970; Krutetskii, 1976; Menchinskaya, 1959; Shapiro, 1965; Yakiman-skaya, 1970).

Finally, the gifted students exhibited great tenacity and perseverance (Burton, 1984; Diezmann & Watters, 2003; Sriraman, 2003a) and stuck with the problem until they were absolutely convinced about their conclusion. The gifted student's approach to proof in this study was very different from the logical approach found in proof in most textbooks and very similar to those used by professional mathematicians. The processes used by the gifted students to prove the truth of the student showed remarkable iso-morphisms to those employed by professional mathematicians as discussed in the previous section.

The logical approach is an artificial reconstruction of discoveries that are being forced into a deductive system and in this process the intuition that guided the discovery process gets lost. The implication here is that many teachers use the logical approach to proof in the classroom and thereby subdue the gifted student's intuition and natural ways of thinking about the problem. These four students would eventually encounter the study of geometry from an intuitive standpoint in the second year of the research based and standards aligned Integrated mathematics course, in which geometry is introduced from an inductive and intuitive standpoint in the context of transformations. In this sequence, the necessity for formal proof is gradually introduced. However gifted students enrolled in a traditional sequence of mathematics courses encounter the study of Euclidean geometry and deductive proof, which robs them of using their natural instincts for establishing truth like mathematicians do. The implication for gifted education is to develop mathematics curricula that creates opportunities for gifted students to develop their intuition about proof and makes use of challenging and worthwhile mathematical tasks.

LIMITATIONS

The students in this study were freshman enrolled in various sections of integrated math in a rural high school. Demographically speaking they were all white, with middle class socioeconomic backgrounds. They all the same K–8 educational background. All four students had very high academic aspirations and intended taking Integrated Math 4 and AP Calculus concurrently in their senior year. The four students had a positive disposition towards mathematics and had enjoyed a high degree of success on all previous mathematical endeavors. These students had not been exposed to constructing mathematical proofs, nor had they formally studied geometry. The results of this study are attributable to the unique characteristics of the population, the particular problem chosen and interview design. The processes used by the mathematically gifted students to construct a proof and their intuitive notions of proof showed remarkable similarities to those of professional mathematicians. In order to generalize these findings to mathematically gifted students that are entering high school with similar middle school backgrounds more research is needed at the early high school level. It is certainly feasible to replicate this experiment with similar types of open ended problems that call for establishing the truth or falsity of mathematical statements.

The author conjectures that mathematically gifted students have the natural intuitive dispositions of mathematicians. It would be a worthwhile endeavor for the gifted education community to have a deeper understanding of these dispositions in order to develop high school curriculum and pedagogy that nurture these natural talents.

REFERENCES

Almeida, D. (2003) Numerical and proof methods of Indian mathematics for the classroom. *Mathematics in School, 32*(2), 7–10.

Bell, A. W. (1976). A study of pupils' proof explanations in mathematical situations. *Educational Studies in Mathematics, 7*, 23–40.

Birkhoff, G. (1969). Mathematics and psychology. *SIAM Review, 11*, 429–469.

Burton, L. (1984). Mathematical thinking: The struggle for meaning. *Journal for Research in Mathematics Education, 15*, 35–49.

Burton, L. (1999). The practices of mathematicians: What do they tell us about coming to know mathematics? *Educational Studies in Mathematics, 37*(2), 121–143.

Casey, E. S. (1978). *Imaging: A phenomenological study.* Penguin Books.

Chang, L. L. (1985). Who are the mathematically gifted elementary school children? *Roeper Review, 8*(2), 76–79.

Chazan, D. (1993). High school geometry students' justification for their views of empirical evidence and mathematical proof. *Educational Studies in Mathematics, 24*, 359–387.

Corbin, J., & Strauss, A. (1998). *Basics of qualitative research.* Thousand Oaks, CA: Sage.

Diezmann, C., & Watters, J. (2003). The importance of challenging tasks for mathematically gifted students. *Gifted and Talented International, 17*(2), 76–84.

Davis, P. J., & Hersh, R. (1981). *The mathematical experience.* New York: Houghton Mifflin.

Dubinsky, E. (1991). Constructive aspects of reflective abstraction in advanced mathematics. In L. P. Steffe (Ed.) *Epistemological foundations of mathematical experience* (pp. 160–187). New York: Springer-Verlag.

Epp, S. S.(1990): The role of proof in problem solving. In A. Schoenfeld (Ed.), *Mathematical thinking and problem solving.* (pp. 257–269). Hillsdale, NJ: Lawrence Erlbaum Associates.

Ernest, P. (1991). *The philosophy of mathematics education.* The Falmer Press.

Ervynck, G. (1991). Mathematical creativity. In D. Tall (Ed.). *Advanced mathematical thinking* (pp. 42–53). The Netherlands: Kluwer Academic Publishers..

Fawcett, H. P. (1938). *The nature of proof.* New York: Teachers College, Columbia University.

Fischbein, E. (1980, August). *Intuition and proof.* Paper presented at the 4th conference of the International Group for the Psychology of Mathematics Education, Berkeley, CA.

Frensch, P., & Sternberg, R. (1992). *Complex problem solving: Principles and mechanisms.* Mahwah, NJ: Erlbaum.

Glaser, B., & Strauss, A. (1977). *The discovery of grounded theory: Strategies for qualitative research.* San Francisco: University of California San Francisco.

Goldberg, A., & Suppes, P. (1972). A computer assisted instruction program for exercises on finding axioms. *Educational Studies in Mathematics, 4,* 429–449.

Greenes, C. (1981). Identifying the gifted student in mathematics. *Arithmetic Teacher, 28*(6), 14–17.

Hadamard, J. W. (1945). *Essay on the psychology of invention in the mathematical field.* Princeton University Press.

Held, M. K. (1983). Characteristics and special needs of the gifted student in mathematics. *The Mathematics Teacher, 76,* 221–226.

Hersh, R. (1993). Proof is convincing and explaining. *Educational Studies in Mathematics, 24,* 389–399.

Hershkowitz, R. (1989). Visualization in geometry—two sides of the coin. *Focus on Learning Problems in Mathematics, 11,* 61–76.

Hoyles, C. (1997). The curricular shaping of students' approaches to proof. *For the Learning of Mathematics, 17*(1), 7–16.

Inhelder, B. & Piaget, J. (1971). *Mental imagery in the child.* Basic Books Inc.

Ivanitsyna, E. N. (1970). Achieving skill in solving geometry problems. In J. Kilpatrick & I. Wirszup (Eds.), *Soviet Studies in the Psychology of Learning and Teaching Mathematics (Vol. 4).* Stanford: School Mathematics Study Group.

Johnson, M. L. (1983). Identifying and teaching mathematically gifted elementary school children. *Arithmetic Teacher, 30*(5), 25–26; 55–56.

Kamii, C., & DeClark, G. (1985). *Young children re-invent arithmetic: Implications of Piaget's theory.* New York: Teachers College Press, Columbia University.

Kanevsky, L. S. (1990). Pursuing qualitative differences in the flexible use of a problem solving strategy by young children. *Journal for the Education of the Gifted, 13,* 115–140.

Kline, M. (1976). NACOME: Implications for curriculum design. *Mathematics Teacher, 69,* 449–454.

Krutetskii, V. A.(1976). *The psychology of mathematical abilities in school children.* (J. Teller, trans. and J. Kilpatrick & I. Wirszup, Eds.). Chicago: University of Chicago Press.

Lakatos, I. (1976). *Proofs and refutations.* Cambridge, UK: Cambridge University Press.

Lampert, M. (1990). When the problem is not the question and the solution is not the answer: Mathematical knowing and teaching. *American Educational Research Journal, 27,* 29–63.

Lerman, S. (1983). Problem-solving or knowledge centered: The influence of philosophy on mathematics teaching. *International Journal of Mathematics Education, 14*(1), 59–66.

Manin, Y. I.(1977). *A course in mathematical logic,* New York: Springer-Verlag.

Mason, J. (1996). Expressing generality and roots of algebra. In N. Bednarz, C. Kieran, &L. Lee (Eds.), *Approaches to algebra* (pp. 65–86). The Netherlands: Kluwer Academic Publishers.

Mason, J., Burton, L., & Stacey, K.(1982). *Thinking mathematically.* London: Addison-Wesley.

Menchinskaya, N. A. (1959). *Psychology of the mastery of knowledge in school.* Moscow: APN Press.

National Council of Teachers of Mathematics. (2000). *Principles and standards for school mathematics.* Reston, VA: Author.

Piaget, J. (1975). *The child's conception of the world.* Totowa, NJ: Littlefield, Adams.

Poincaré, H. (1948). *Science and method.* New York: Dover.

Polya, G. (1954). *Mathematics and plausible reasoning: Induction and analogy in mathematics* (Vol.1). Princeton, NJ: Princeton University Press.

Presmeg, N. C. (1986). Visualization and mathematical giftedness. *Educational Studies in Mathematics, 17,* 297–311.

Rubin, A., & Babbie. E. (1997) *Research Methods for Social Work* (3rd ed.), Pacific Grove, CA: Brooks/Cole Publishing Company.

Senk, S. (1985). How well do students write geometry proofs? *Mathematics Teacher, 78,* 448–456.

Shapiro, S. I. (1965). A study of pupil's individual characteristics in processing mathematical information. *Voprosy Psikhologii,* No. 2.

Sheffield, L. J. (1999). *Developing mathematically promising students.* Reston, VA: National Council of Teachers of Mathematics.

Sriraman, B. (2002). How do mathematically gifted students abstract and generalize mathematical concepts. *NAGC 2002 Research Briefs, 16,* 83–87.

Sriraman, B. (2003a). Mathematical giftedness, problem solving, and the ability to formulate generalizations. *The Journal of Secondary Gifted Education, XIV*(3), 151–165.

Sriraman, B. (2004). The characteristics of mathematical creativity. *The Mathematics Educator, 14*(1), 19–34.

Strunz, K. (1962). *Pädogogische Psychologie des mathematischen Denkens.* Heidelberg: Quelle & Meyer.

Suppes, P., & Binford, F. (1965). Experimental teaching of mathematical logic in the elementary school. *The Arithmetic Teacher, 12,* 187–195.

Usiskin, Z. P. (1987). Resolving the continuing dilemmas in school geometry. In M. M. Lindquist, & A. P. Shulte (Eds.) *Learning and teaching geometry, K–12: 1987 Yearbook* (pp. 17–31). Reston, VA: National Council of Teachers of Mathematics.

Van Hiele, P. M. (1986). *Structure and insight.* Orlando, FL: Academic Press.

Wallas, G. (1926). *The art of thought.* New York: Harcourt Brace.

Wertheimer, M. (1945). *Productive thinking.* New York: Harper.

Yakimanskaya, I. S. (1970). Individual differences in solving geometry problems on proof. In J. Kilpatrick & I. Wirszup (Eds.). *Soviet Studies in the Psychology of Learning and Teaching Mathematics (Vol. 4)*, Stanford: School Mathematics Study Group.

ACKNOWLEDGMENT

Reprint of Sriraman, B. (2004). Gifted ninth graders' notions of proof. Investigating parallels in approaches of mathematically gifted students and professional mathematicians. *Journal for the Education of the Gifted, 27*(4), 267–292. Reprinted with permission from Prufrock Press. ©2004 Bharath Sriraman.

NOTE

1. Sarah is referring to exercises in ratio and proportion that give a sequence of steps to establish basic identities.

CHAPTER 4

ARE MATHEMATICAL GIFTEDNESS AND MATHEMATICAL CREATIVITY SYNONYMS?

A Theoretical Analysis of Constructs

Bharath Sriraman
The University of Montana

ABSTRACT

At the K–12 level one assumes that mathematically gifted students identified by out-of-level testing are also creative in their work. In professional mathematics, "creative" mathematicians constitute a very small subset within the field. At this level mathematical giftedness does not necessarily imply creativity but the converse is certainly true. In the domain of mathematics are the words creativity and giftedness synonyms? In this article the constructs of mathematical creativity and mathematical giftedness are developed via a synthesis and analysis of the general literature on creativity and giftedness. The notions of creativity and giftedness at the K–12 and professional levels are compared and contrasted to develop principles and models that theoretical-

Creativity, Giftedness, and Talent Development in Mathematics, pages 85–112

ly "maximize" the compatibility of these constructs. The relevance of these models for the K–12 level and professional levels are discussed with practical considerations for the classroom. The paper also significantly extends ideas presented by Usiskin (2000).

INTRODUCTION

Creativity in mathematics is often looked at as the exclusive domain of professional mathematicians. The word creativity is "fuzzy" and lends itself to a variety of interpretations. What does creativity mean in mathematics? Is it purely the discovery of an original result? If so, then creativity is indeed the exclusive domain of professional mathematicians. Does student discovery of a hitherto known result or an innovative mathematical strategy also constitute creativity? Eminent mathematicians like Jacques Hadamard (1945) and George Polya (1954) have said that the only difference between the work of a mathematician and a student is that of degree. In other words, each operates at their respective levels and we should recognize that students are also capable of being creative. Such a view is especially relevant to teachers of mathematically gifted students, who would expect gifted students to display creative traits. Does being mathematically gifted students pre-disposes students to being creative? In other words, if a student has been identified as being mathematically gifted, then are they also creative in their approach to mathematics? Does mathematical giftedness imply mathematical creativity?

Kajander (1990) has stated that even among the mathematically gifted who displayed creative traits such as divergent thinking, mathematical creativity was a "special kind of creativity not necessarily related to divergent thinking" (p. 254). This statement begs the question as whether mathematical creativity implies giftedness. The statement is certainly true at research level mathematics. One could easily argue that professional mathematicians are gifted based on the fact that they have obtained a doctorate in the field and are active in research. However even at this level, professional mathematicians classify only a small handful of their colleagues as being truly "creative" (Usiskin, 2000).

An examination of Usiskin's (2000) eight-tiered hierarchy may help clarify the degrees of giftedness and creativity wih regard to mathematics. Usiskin devised this hierarchy to classify mathematical talent, which ranges from Level 0 to Level 7. In this hierarchy Level 0 (No Talent) represents adults who know very little mathematics, Level 1 (Culture level) represents adults who have rudimentary number sense as a function of cultural usage and their mathematical knowledge is comparable to those of students in grades 6–9. It is obvious that a very large proportion of the general population would fall into the first two levels.

Thus, the remaining population is thinly spread out into levels 2 through 7 on the basis of mathematical talent. Level 2 represents the honors high school student who is capable of majoring in mathematics as well as those that eventually become secondary math teachers. Level 3 (the "terrific" student) represents students that score[1] in the 750–800 on the SAT's or 4 or 5 in the Calculus AP exams. These students have the potential to do beginning graduate level work in mathematics. Level 4 (the "exceptional" student) represents students that excel in math competitions and receive admission into math/science summer camps and/or academies because of their talent. This student is capable of constructing mathematical proofs and able to "converse" with mathematicians about mathematics. Level 5 represents the productive mathematician. Although Usiskin's (2000) description of this level is vague, one can infer that it represents a student that has successfully completed a Ph.D. in the mathematics or related mathematical sciences and is capable of publishing in the field. Level 6 is the rarified territory of the exceptional mathematician, which represents "mathematicians…[that] have moved their domains forward with notable conquests; they will be found in any history of the domains in which they work. These are at the level of the Alfred P. Sloan fellows, the best in their age group in the country." (Usiskin, 2000). Finally Level 7 are the all-time greats, the Fields medal winners in mathematics.[2] This level is the exclusive territory of giants or exemplary geniuses like Leonard Euler, Karl Friedrich Gauss, Bernhard Riemmann, Srinivasa Ramanujan, David Hilbert, Henri Poincaré, and others.

In Usiskin's (2000) eight-tiered hierarchy of mathematical talent, the professional (gifted) mathematician is at level 5 whereas the creative mathematician is found at levels 6 and 7. Therefore in the professional realm mathematical creativity implies mathematical giftedness but the reverse is not necessarily true. In this hierarchical classification of mathematical talent, students that are gifted and/or creative in mathematics are found at levels 3 and 4. The point that Usiskin (2000) emphasized is that these students have the potential of moving up into the professional realm (level 5) with appropriate affective and instructional scaffolding as they progress beyond the K–12 realm into the university setting.

MOTIVATION FOR THIS PAPER

Numerous studies (i.e., Cramond, 1994; Davis, 1997; Smith, 1966; Torrance, 1981) indicate that the behavioral traits of creative individuals very often go against the grain of acceptable behavior in the institutionalized school setting. For instance, negative behavioral traits such as indifference to class rules, display of boredom, cynicism or hyperactivity usually result in disciplinary measures as opposed to appropriate affective interventions. In the case of gifted students who "conform" to the norm these students are often prone to hide their intellectual capacity for social reasons, and identify their academic talent as being a source of envy (Massé & Gagné, 2002). History is peppered with numerous examples of creative individuals described as "deviants" by the status quo. Brower (1999) has presented over fifty examples of eminent writers, moral innovators, scientists, artists and stage performers that were jailed because of society's fear of ideas that are "out there" but powerful enough to create paradigmatic shifts in the public's mindset.

The stifling of creativity at the K–12 levels is often collectively rationalized under the guise of doing what is supposedly good for the majority of the students, invoking the oft-misused term "equity," appealing to curricular plans and school achievement goals, and so forth. The recent passing of the No Child Left Behind Act (NCLB) in the United States under the guise of "equity" has brought to the forefront the debate of what is to be done with creative and gifted students in the classroom. Recently Marshak (2003) wrote that the NCLB call for accountability based on standardized testing for the traditional skills of reading, writing and arithmetic valued in the industrial societal setting is a giant step backwards to the 1940's. Based on recent reports released by the U.S. Department of Labor, Marshak (2003) has further stated that aside from the three "traditional" R's (reading, writing, and 'rithmatic), numerous additional skills such as problem solving and creative thinking skills are necessary for success in the global societal setting of 21st century. Even at the tertiary levels there have been criticisms about the excessive amount of structure imposed on disciplines by academics as well as the "narrow, profoundly western centric attitudes" (Creme, 2003, p. 273). Such a criticism particularly resonates in the world of mathematics, especially at the K–12 level, where gifted/creative students with a non-western ethnic background are rarely encouraged to express or use perfectly reasonable mathematical techniques they may be familiar with from their own cultures. They are instead taught to adopt a western attitude. In summary, the literature indicates that giftedness is often associated with conformity whereas creativity is viewed as a fringe commodity, tolerated and nurtured by some teachers but typically not encouraged. There is clearly a schism between the value of creativity in the K–12 and professional realms, which leads one to ponder whether and/or how this schism can be bridged. This question is further explored in the present paper, *What about problem solving?*

Problems tackled by professional mathematicians are full of uncertainty. However, most curricular and pedagogical approaches rarely offer students this open-ended view of mathematics. In fact classroom practices and math curricula rarely use ill-posed or open-ended problems nor do they allow students a prolonged period of engagement and independence to work on these problems.

Encouragingly, problem solving in the mathematics classroom has received increased emphasis since the release of the original National Council of Teachers of Mathematics Standards (1989). However, nearly two decades later, it has essentially become a dogmatic term invoked to act as panacea to remedy curricular ills. This rather strong statement receives considerable support from the extant surveys of the research literature on problem solving. For example, in the *Handbook for Research on Mathematics Teaching and Learning*, Schoenfeld (1993) described how the field of mathematics education in the United States has been subject to approximately 10-year cycles of pendulum swings between basic skills and problem solving. He concluded his chapter with optimism about the continuation of a movement that many at that time referred to as "the decade of problem solving" in mathematics education. However, since the 1993 handbook was published "the worldwide emphasis on high-stakes testing has ushered in an especially virulent decade-long return to basic skills" (Lesh & Sriraman, 2005b, p. 501).

Additionally, consider the following facts: Polya-style problem solving heuristics—such as *draw a picture, work backwards, look for a similar problem,* or *identify the givens and goals*—have long histories of being advocated as important abilities for students to develop (Polya, 1945). But, what does it mean to "understand' them? Such strategies clearly have descriptive power. That is, experts often use such terms when they give after-the-fact explanations of their own problem solving behaviors—or those of other people that they observe. But, there is little evidence that general processes that experts use to describe their past problem solving behaviors should also serve well as prescriptions to guide novices' next steps during ongoing problem solving sessions. Researchers gathering data on problem solving also have the natural tendency to examine the data in front of them through the lens of a prioi problem solving models. Although there is great value in doing so, does such an approach really move problem-solving research forward? If one examines the history of problem solving research, there have been momentous occasions when researchers have realized the restricted "heuristic" view of problem solving offered by the existing problem solving research "toolkits" and have succeeded in re-designing existing models with more descriptive processes. However the problem remains that descriptive processes are really more like names for large categories of skills rather than being well defined skills in themselves. Therefore, in attempts to go beyond "descriptive power" to make such processes more "prescriptive power", one tactic that researchers and teachers have attempted is to convert each "descriptive process" into longer lists of more-restricted-but-also-more-clearly-specified processes. However, if this approach is adopted, most of what it means to understand such processes involves knowing when to use them. So, "higher order" managerial rules and beliefs need to be introduced which specify when and why to use "lower order" prescriptive processes.

The obvious dilemma that arises is that short lists of descriptive processes have appeared to be too general to be meaningful. On the other hand, long lists of prescriptive processes tend to become so numerous that knowing when to use them becomes the heart of understanding them (Lesh & Sriraman, 2005b). Furthermore, adding more meta-cognitive rules and beliefs only compounds these two basic difficulties. A decade after Schoenfeld's (1993) *Handbook for Research on Mathematics Teaching and Learning* was published, in another extensive review of the literature, Lester and Kehle (2003) again reported that little progress had been made in problem solving research—and that problem solving still had little to offer to school practice. Their conclusions agreed with Silver (1985), who long ago determined what we consider to be the core of the problem in problem solving research. That is, the field of mathematics education needs to go "beyond process-sequence strings and coded protocols" in our research methodologies and "simple procedure-based computer models of performance" to develop ways of describing problem solving in terms of conceptual systems that influence students' performance (Silver, 1985, p. 257). Thus, the use of problem solving in the mathematics classroom arguably lends itself to a host of questions about its purpose as well as its effectiveness.

MATHEMATICAL CREATIVITY: THE LACK OF DOMAIN SPECIFIC DEFINITIONS IN MATHEMATICS

Given the preamble of giftedness and creativity and its relevance in our society, we now focus our attention more specifically and deeply into to the domain of mathematics with the purpose of generating appropriate definitions for these terms. The existing literature is used to survey the various meanings of the terms mathematical creativity and mathematical giftedness and to determine their compatibility and relevance at the professional and K–12 levels. Most of the extant definitions of mathematical creativity found in the mathematics and mathematics education literature is vague or elusive. This ambiguity may exist because of the difficulty of describing this complex construct. For instance mathematical creativity has been defined via the use of various metaphors such as the ability to discern or choose (Hadamard, 1945; Poincaré, 1948), to distinguish between acceptable and unacceptable patterns (Birkhoff, 1969), and to engage in non-algorithmic decision-making (Ervynck, 1991). The literature on students that are mathematically creative in the K–12 realm is also vague. Exceptional mathematical ability (Level 4 talent) in the K–12 realms has been associated with the Einstein syndrome (Sowell, 2001) and the Asperger syndrome (Jackson, 2002). The Einstein syndrome is characterized by exceptional mathematical ability but delayed speech development, whereas the Asperger syndrome is a spectrum disorder characterized by "severe impairment in reciprocal social interaction, all absorbing narrow interests or obsession with a particular subject… [a]nd sometimes motor clumsiness" (James, 2003, p. 62). The dearth of specific definitions on mathematical creativity in the mathematics and mathematics education literature, necessitates that we move away from the specific domain of mathematics to the general literature on creativity with the goal of constructing an appropriate definition.

Creativity: General Definitions in Psychology/Education Psychology

In the general literature on creativity numerous definitions can be found. Craft (2002) used the term "life wide creativity" to describe the numerous contexts of day to day life in which the phenomenon of creativity manifests. Other researchers have described creativity as a natural "survival" or "adaptive" response of humans in an ever-changing environment (Gruber, 1989; Ripple, 1989). Craft (2003) has pointed out that it is essential we distinguish "everyday creativity" such as improvising on a recipe from "extraordinary creativity" which causes paradigm shifts in a specific body of knowledge. It is generally accepted that works of "extraordinary creativity" can be judged only by experts within a specific domain of knowledge (Csikszentmihalyi, 1988, 2000; Craft, 2003). For instance, Andrew Wiles' proof of Fermat's Last Theorem could only be judged by a handful of mathematicians within a very specific sub-domain of number theory.

More specifically, the the realm of educational psychology, one can also find a variety of definitions of creativity. For example, Weisberg (1993) suggested that creativity entails the use of ordinary cognitive processes and results in original and extraordinary products. Further, Sternberg and Lubart (2000) defined creativity as the ability to produce unexpected original work, which is useful and adaptive. Other definitions usually impose the requirement of novelty, innovation or unusualness of a response to a given problem (Torrance, 1974). Numerous confluence theories of creativity define creativity as a convergence of knowledge, ability, thinking style, motivational and environmental variables (Sternberg & Lubart, 1996, 2000), an evolution of domain specific ideas resulting in a creative outcome (Gruber

& Wallace, 2000). For example, Csikzentmihalyi (2000) suggests creativity is one of mutations resulting from a favorable interaction between an individual, domain and field. Most recently, Plucker and Beghetto (2004) offered an empirical definition of creativity based on a survey and synthesis of numerous empirical studies in the field. They defined creativity as "the interplay between ability and process by which an individual or group produces an outcome or product that is both novel and useful as defined within some social context" (p. 156).

Applying the General Definitions of Creativity to Mathematics

At the K–12 level, one normally does not expect works of extraordinary creativity, however it is certainly feasible for students to offer new insights into a math problem or a new interpretation or commentary to a literary or historical work. Students at the K–12 level are certainly capable of originality. A synthesis of the numerous definitions of creativity can lead to a working definition of mathematical creativity at both the professional and K–12 levels. At the professional level mathematical creativity can be defined as (a) the ability to produce original work that significantly extends the body of knowledge, and/or (b) one who opens up avenues of new questions for other mathematicians.

For instance Hewitt's (1948) paper on rings of continuous functions led to unexplored possibilities and questions in the fields of analysis and topology which sustained other mathematicians for decades. A modern day illustration on the far-reaching effects of Hewitt's paper is to "google" the title of the paper, which results in over 120,000 hits.

On the other hand, mathematical creativity at the K–12 can be defined as (a) the process that results in unusual (novel) and/or insightful solution(s) to a given problem or analogous problems, and/or (b) the formulation of new questions and/or possibilities that allow an old problem to be regarded from a new angle requiring imagination (Einstein & Inheld, 1938; Kuhn, 1962). The second part of this definition is very similar to those for creativity in professional mathematics.

The research also indicates that at both the K–12 and the professional levels creative individuals are prone to reformulating a problem or finding analogous problems (Polya, 1945, 1954; Frensch & Sternberg, 1992). They are also different from their peers in that they are fiercely independent thinkers (Chambers, 1964; Gruber, 1981;Ypma, 1968), tend to persevere (Chambers, 1964; Diezmann & Watters, 2003) and to reflect a great deal (Policastro & Gardner, 2000; Sriraman, 2003, Wertheimer, 1945).

Conditions that Enhance Mathematical Creativity at the Professional Level

Now that we have a working definition of mathematical creativity, it is natural to explore the conditions under which such creativity manifests. In order to illuminate the conditions that enhance the manifestation of creativity at the professional level, Sriraman (2004c) conducted a qualitative study with 5 accomplished and creative professional mathematicians, the goal of which was to better understand the conditions under which mathematical creativity manifests. The five mathematicians verbally reflected on the thought processes involved in creating mathematics. The results indicated that in general, the mathematicians' creative process followed the four-stage Gestalt model (Wallas, 1926) of *preparation-incubation-illumination-verification*. It was also found that social interaction, imagery, heuristics, and intuition were some of the characteristics of mathematical creativity. Other characteristics that contributed to their research productivity were the time available in an academic setting for the pursuit of research, freedom of movement, the aesthetic appeal of mathematics and the urgency/drive to solve problems with tremendous real world implications. All five mathematicians spoke at length about the "Aha" or "Eureka" moment (Burton, 1999a, 1999b; Wallas, 1926) at which they gained a new insight into the problem, which led them to successfully construct a proof.

MATHEMATICAL GIFTEDNESS

A synthesis of the research literature on mathematical giftedness and features of mathematical thinking revealed that the construct of mathematical giftedness has been defined in terms of the individual's ability in mathematical processes such as: (a) the ability to abstract, generalize and discern mathematical structures (Kanevsky, 1990; Kiesswetter, 1985, 1992; Krutetskii, 1976; Shapiro, 1965; Sriraman, 2002, 2003); (b) the ability to manage data (Greenes, 1981; Yakimanskaya, 1970); (c) the ability to master principles of logical thinking and inference (Suppes & Binford, 1965; Goldberg & Suppes, 1972); (d) the ability to think analogically and heuristically and to pose related problems (Polya, 1954; Kiesswetter, 1985); (e) flexibility and reversibility of mathematical operations and thought (Krutetskii, 1976); (f) an intuitive awareness of mathematical proof (Sriraman, 2004c); (g) the ability to independently discover mathematical principles (Sriraman, 2004a, 2004b); (h) the ability to make decisions in problem solving situations (Frensch & Sternberg, 1992; Schoenfeld, 1985; Sriraman, 2003); (i) the ability to visualize problems and/ or relations (Hershkowitz, 1989; Presmeg, 1986); (j) the ability to infer behaviors that test for truth or falsity of a construct (Wason & Johnson-Laird,

1972); (k) the ability to distinguish between empirical and theoretical principles (Davydov, 1988, 1990; Vygotsky, 1962, 1978); and (l) the ability to think recursively (Kieran & Pirie, 1991; Minsky, 1985).

In addition mathematical giftedness has also been associated with the capacity for learning at a faster pace (Chang, 1985; Heid, 1983). Most of the mathematical processes listed above are cognitive in feature and learned during K–12 schooling experiences. It should be noted that many of these studies involved task-based instruments with specific mathematical concepts/ideas to which students had some exposure. Another important observation is that although many of these traits do play a role and are necessary in the setting of professional mathematics, they are not sufficient for creativity to manifest. In other words, in order to work as a professional mathematician (level 5) and to create new mathematics, some abilities are more crucial than others. In particular decision making, the abilities of abstracting and generalizing, infering, constructing theoretical principles and recursive thinking play an important role in how mathematics is created at the professional level. The processes of infering (Wason & Johnson Laird, 1972), constructing theoretical principles (Davydov, 1988, 1990) and recursive thinking (Kieren & Pirie,1991; Vitale,1989) play a vital role in how new mathematics is created. Simply put this process can be viewed as follows: The applied mathematician is trying to create mathematical models that say something about the physical world. The pure mathematician is willing to take those models and see what the implications are. During the modeling process, there exists some physical situation and the applied mathematician tries to identify the underlying principles. The pure mathematician steps back and abstracts them to have a setting in which these basic principles hold, and to see what the implications are. The pure mathematician who is dealing with the implications is working rather formally to see what is implied logically by this particular set of assumptions, without worrying about whether this is an appropriate model or not.

Wason and Johnson-Laird (1972) investigated whether adults, when given a set of assertions were able to appreciate the logical implications. They were particularly interested in determining contexts that led adults[3] into drawing fallacious conclusions. According to Wason and Johnson-Laired, "The rational individual, in our sense of the word, is merely one who has *the ability to make inferences*; he may not be rational in any other sense of the word." (p. 2). The process of infering can lead to mathematical generalizations. Thus, Wason and Johnson-Laird (1972) investigated how adults discovered general rules, by setting up structured experiments, in which "subjects were presented with a hypothesis and they had to decide the items of evidence relevant for testing its truth. The experiments were designed to investigate the propensity of individuals to offer premature solutions based on confirming evidence" (p. 202).

The researchers had several interesting findings. First of all, subjects tended to make fallacious inferences, when presented with affirmative statements. Another finding was that an over-whelming majority of the subjects in the study were prone to try to verify generalizations rather than to try and falsify them. Moreover, the researchers noted that the content of the material about which inferences are made were significant. Subjects tended to make "illicit conversions" and were biased towards verification, when faced with material of an abstract nature, such as mathematical problems in n-dimensional space which can only be symbolically represented. However, when the material was concrete and subjects had experienced a variety of connections with them, they tended to generate and assess hypothetical connections between facts. In other words, mathematical inference behavior is different from everyday inference behavior and mathematically gifted individuals are adept at logically and correctly connecting abstract constructs different from everyday constructs. This distinction between everyday (or empirical) and abstract (or theoretical) concepts was studied by Vygotsky (1962, 1978) in his investigations on concept formation and later pursued by Davydov (1988, 1990).

Initially Vygotsky (1962, 1978) explored the notion of scientific generalizations during his investigations on concept formation and distinguished between two types of concepts, namely spontaneous or everyday concepts, and scientific or theoretical concepts. Davydov (1988) who continued this line of investigation on conceptualization stressed that the important difference between everyday (empirical) concepts and theoretical concepts lies in their mode of formation. According to Davydov (1988) the difference between everyday concepts and theoretical concepts also lies in the type of abstraction one engages in, namely empirical abstraction versus theoretical abstraction. The former involves superficial comparisons for discerning similarities and differences whereas the latter involves structural comparisons. Thus, empirical generalization requires the abstraction of similarities from collections of entities, which may themselves represent disparate functions and structures. For instance Davydov (1988) said that the notion of "roundness" can be empirically abstracted from a dish, a wheel, and so forth. However, this empirical notion of circularity does not reveal the real objective content, which is the locus of points at a constant distance from a fixed point. This content is not apparent from the mere appearance of roundness. Davydov (1988, 1990) claimed that cultivating empirical generalization useful only for the formation of everyday concepts is inadequate for the formation of theoretical generalizations, which characterize mathematics. Aside from infering behavior in theoretical situations, mathematical thinking is also characterized by "recursive thinking" a term borrowed from information processing.

According to Vitale (1989) recursion is the mode in which human beings tackled and represented problems. Kieren and Pirie (1991) claim that recursion is an appropriate metaphor "in looking at the complex phenomenon of the whole of a person's mathematical knowledge and mathematical understanding." (p. 79). Kieren and Pirie substantiate this claim by arguing that children are self-referencing and in their realm of existence there are a number of "behavioral possibilities." Consequently, because children are self-referencing, one of the primary means of cognition is recursive in nature, and their knowledge "is formed through thought actions which entail the results of previous thought actions as inputs" (p. 79). This aspect of cognition is particularly relevant in mathematics because mathematical knowledge building and understanding is a dynamic process in which one's present knowledge and understanding builds from and is linked to previous knowledge. Kieren and Pirie (1991) analyzed recursion in student's thoughts and actions within a problem solving experience. They posed the well know "handshake problem": How many handshakes are needed in a class with 35 students so that each person in a room shakes hands with every person exactly once?

One of the solution strategies used by a group was to first specialize the problem involving thirty people to one involving the group. The students devised a strategy whereby the people were lined up, and the person furthest from the door shook hands with everybody else, reported the number of handshakes to the last person with whom the handshake occurs, and then to left the room. Then the second person furthest from the door repeated the above procedure, reported the number of handshakes to the last person, and left the room. After many iterations, the last person added up the number of handshakes, namely $34 + 33 + 32 + \ldots + 1$, and left the room. Note that the above solution is easily generalized to the case with n people. One of the findings of this study was that most of the students never bothered to compute a solution for 35 people, since they had come up with a strategy that would generally work.

The fact that the problem was never answered at any level suggests that the students sense that the solution to the problem does not reduce to an answer or result of a special case, but "calls" or uses the structure of that special case. The "structure" of this special case has in it both a substantially correct mathematical idea (a sequence of non-repeating handshakes) and a form by which it can be procedurally described" (Kieren & Pirie, 1991, p. 83–84).

The researchers envisioned this recursive structure diagrammatically as a triangle of activities, which included specializing, creating results, and then generalizing through interpretation and validation. The process of specializing to particular cases, conjecturing and then generalizing through interpretation and validation is a common trait among mathematically gifted students (Krutetskii, 1976; Sriraman, 2003, 2004d). This process also shows similarities to how professional mathematicians interpret and extend results in their field (Sriraman, 2004c).

Discussion, Implications, and Recommendations for the K–12 Classroom

The preceding discussions reveal that although mathematically gifted students possess many of the cognitive qualities required for work at the professional level, some cognitive traits are more important than others. This hierarchy implies a use of problems that call for the use of a high level of inferencing, generation/discovery of principles and recursive thinking.

The discussion on mathematical creativity indicates that many of the characteristics of mathematical creativity described by mathematicians as invaluable aspects of their craft such as the freedom to choose and pursue problems in an academic setting, the freedom of movement required during work, the awareness of the distinction between learning versus creating, the aesthetic appeal of mathematics and the affective urgency/drive to solve problems with tremendous real world implications, might be extremely difficult to simulate in a traditional classroom setting. A model involving the use of five principles to maximize creativity among the mathematically gifted in the K–12 setting is shown in Figure 4.1. I have outlined five general principles extracted from the literature and studies on mathematical

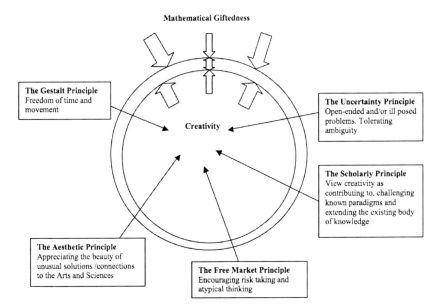

Figure 4.1 Harmonizing creativity and giftedness at the K–12 level.

creativity that can be applied in the everyday classroom setting in order to maximize the potential for mathematical creativity to manifest in the K–12 classroom.

FIVE OVERARCHING PRINCIPLES TO MAXIMIZE CREATIVITY

As seen in Figure 4.1, the five overarching principles that emerged from a synthesis and analysis of the literature as significantly enhancing mathematical creativity are labeled as (a) the Gestalt principle, (b) the Aesthetic principle, (c) the free market principle, (d) the scholarly principle, and (e) the uncertainty principle.

The Gestalt principle. The eminent French mathematicians Hadamard (1945) and Poincaré (1948) viewed creativity as a process by which the mathematician makes choices between questions that lead to fruition as opposed to those that lead to nothing new. These mathematicians were influenced by the Gestalt psychology of their time and characterized mathematical creativity as a four-stage process consisting of preparation, incubation, illumination, and verification (Wallas, 1926). Although psychologists have criticized the Gestalt model of creativity because it attributes a large "unknown" part of creativity to unconscious drives during incubation, numerous studies with scientists and mathematicians (i.e. Burton, 1999a, 1999b; Davis & Hersh, 1981; Shaw, 1994; Sriraman, 2004c) have consistently validated this model. In all these studies after one has worked on a problem for a considerable time (preparation) without making a breakthrough, one puts the problem aside and other interests occupy the mind. This period of incubation eventually leads to an insight on the problem, to the "Eureka" or the "Aha!" moment of illumination. Most of us have experienced this magical moment. Yet the value of this archaic Gestalt construct is ignored in the classroom. In fact Krutetskii (1976) found that mathematically gifted children also experienced the sheer joy of creating that "included the feeling of satisfaction from the awareness of the difficulties that have been overcome, that one's own efforts have led to the goal" (p. 347). This implies that it is important that teachers encourage the mathematically gifted to engage in suitably challenging problems over a protracted time period thereby creating the opportunities for the discovery of an insight and to experience the euphoria of the "Aha!" moment.

The aesthetic principle. Mathematicians have often reported the aesthetic appeal of creating a "beautiful" theorem that ties together seemingly disparate ideas, combines ideas from different areas of mathematics, or utilizes an atypical proof technique (Birkhoff, 1956, 1969; Dreyfus & Eisenburg, 1986; Hardy, 1940). Wedderburn's theorem that a finite division ring is a field is one instance of a unification of apparently random ideas because the proof involves algebra, complex analysis and number theory. Cantor's argument about the uncountability of the set of real numbers is an oft quoted example of a brilliant and atypical mathematical proof technique (Nickerson, 2000). The eminent English mathematician G.H. Hardy (1940) compared the professional mathematician to an artist, because like an artist, a mathematician was a maker of patterns in the realm of abstract ideas. Hardy (1940) said:

> A mathematician, like a painter or a poet, is a maker of patterns. If his patterns are more permanent than theirs, it is because they are made with ideas....The mathematician's patterns, like the painter's or the poet's, must be beautiful; the ideas, like the colors or the words, must fit together in a harmonious way. Beauty is the first test: there is no permanent place in the world for ugly mathematics. (p. 13)

Recent studies in Australia (Barnes, 2000), and Germany (Brinkmann, 2004) with middle and high school students revealed that students were capable of appreciating the "aesthetic beauty" of a simple solution to a complex mathematical problem. Brinkmann (2004) found that even low achievers appreciated the struggle to get the insight that unlocked a seemingly unsolvable mathematical puzzle. Barnes (2000) recommended real world problem selection and the careful "staging" of the discovery moment by the teacher were found to be the crucial elements for conveying an appreciation of mathematics to the classroom.

The free market principle. Professional mathematicians in an academic setting take a huge risk when they announce a proof to a long standing unsolved problem. Often times the mathematician puts their reputation at risk if a major flaw is discovered in their proof. For instance in mathematical folklore, Louis De Branges' announcement of a proof to the Riemann hypothesis[4] fell through upon scrutiny by the experts. This led to subsequent ignorance of his claim to a brilliant proof for the Bieberbach Conjecture.[5] The western mathematical community took notice of Louis De Brange's proof of the Bieberbach conjecture only after a prominent Soviet group of mathematicians supported his proof. On the other hand, Ramanujan's numerous intuitive claims, which lacked proof, were widely accepted by the community because of the backing of giants like G.H. Hardy and J.E. Littlewood. The implication of these anecdotes from professional mathematics for the classroom is that teachers should encourage students to take risks. In particular they should encourage the gifted/creative students to pursue and present their solutions to contest or open problems at appropriate regional and state math student meetings, allowing them to gain experience at defending their ideas upon scrutiny from their peers.

The scholarly principle. K–12 teachers should embrace the idea of "creative deviance" as contributing to the body of mathematical knowledge, and they should be flexible and open to alternative student approaches to problems. In addition, they should nurture a classroom environment in which students are encouraged to debate and question the validity of both the teachers' as well as other students' approaches to problems. Gifted students should also be encouraged to generalize the problem and/or the solution as well as pose a class of analogous problems in other contexts. Allowing students problem posing opportunities and understanding of problem design helps them to differentiate mathematical problems from non-mathematical problems, good problems from poor, and solvable from non-solvable problems. In addition, independent thinking can be cultivated by offering students the opportunity to explore problem situations without any explicit instruction (English, in press; Sriraman & English, 2004). Teachers are also encouraged to engage in curriculum acceleration and compaction to lead mathematically gifted students into advanced concepts quickly to promote independent scholarly activity. The longitudinal Study of Mathematically Precocious Youth (SMPY) started by Julian Stanley at Johns Hopkins in 1971 generated a vast amount of empirical data gathered over the last 30 years, and has resulted in many findings about the types of curricular and affective interventions that foster the pursuit of advanced coursework in mathematics. More than 250 papers have been produced in its wake, and they provide excellent empirical support for the effectiveness of curriculum acceleration and compaction in mathematics (Benbow, Lubinski, & Sushy, 1996).

The uncertainty principle. Mathematics at the professional level is full of uncertainty and ambiguity as indicated in some of the quotes presented earlier. Creating, as opposed to learning, requires that students be exposed to the uncertainty as well as the difficulty of creating mathematics. This ability requires the teacher to provide affective support to students who experience frustration over being unable to solve a difficult problem. Students should periodically be exposed to ideas from the history of mathematics and science that evolved over centuries and took the efforts of generations of mathematicians to finally solve. Cultivating this trait will ultimately serve the mathematically gifted student in the professional realm. Keisswetter (1992) developed the so-called *Hamburg Model* in Germany, which is more focused on allowing gifted students to engage in problem posing activities, followed by time for exploring viable and non-viable strategies to solve the posed problems. This approach captures an essence of the nature of professional mathematics, where the most difficult task is to often to correctly formulate the problem (theorem). Conversely, some extant models within the U.S., such as those used in the Center for Talented Youth (CTY), tend to focus on accelerating the learning of concepts and processes from the regular curriculum, thus preparing students for advanced coursework within mathematics (Barnett & Corazza, 1993).

Having presented 5 principles that can maximize the creativity in the K–12 classroom, I present a model (Figure 4.2), which captures the underlying essence of this paper and shows the relationship and compatibility of the constructs of mathematical creativity and giftedness between the K–12 and the professional realms of mathematics.

The Conceptual Model

The model presented in Figure 4.2 shows the dynamic nature of the relationship between mathematical creativity and mathematical giftedness and illustrates the possibilities for bridging the schism between the K–12 and the professional realms of mathematics. The K–12 world in the figure shows

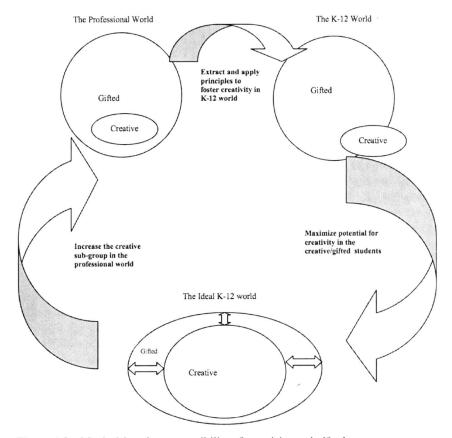

Figure 4.2   Maximizing the compatibility of creativity and giftedness.

that mathematical creativity manifests in the "fringes" in the general pool of mathematically gifted students. On the other hand, the professional world of mathematics shows that creativity is a rare and sought after commodity. How do we bridge these two disjointed worlds? The model suggests that these seemingly disjoint realms of professional mathematics and the K–12 mathematics classroom can be successfully bridged by paying increased attention to maximizing the creative potential of the mathematically gifted students in the "ideal" classroom. This can be accomplished by applying the five principles (Figure 4.1) that work for creative mathematicians in the K–12 classroom. A classroom environment, especially classrooms with mathematically gifted students, in which the Gestalt, aesthetic, free market, scholarly and uncertainty principles become a part of the classroom culture maximizes the potential for creativity among the mathematically gifted. This increase in creativity serves these students as they progress into the world of post secondary and research mathematics, or progress from levels 4 to levels 5 and 6. Progress into level 6 increases the subgroup of creative mathematicians.

The conceptual model although "triadic" in nature is different from the general models proposed by Renzulli (1978, 1986) and Sternberg (1997) in that it shows the relationship between creativity and giftedness in the specific domain of mathematics. However, the model has elements of Renzulli's (1978, 1986) three-ringed conception of giftedness as well as Sternberg's (1997) triachic view of giftedness. Renzulli's (1978) three–ringed conception suggests that giftedness is the interaction between above-average abilities, focused task-committed behavior and creativity. The professional world of mathematics (level 5) is characterized by above-average mathematical abilities among mathematicians and a commitment to research. However mathematical creativity at this level remains an elusive commodity manifesting among a tiny subset of the general pool of mathematicians. In the ideal K–12 world, Renzulli's (1978) conception of task-commitment can be emphasized under the uncertainty principle, which suggests difficult problems take extended time periods to solve and involve considerable struggle. Sternberg's (1997) triarchic view of giftedness suggests that gifted individuals possess a varying blend of analytic, synthetic (creative), and practical giftedness. This view particularly resonates in the world of mathematics. Level 5 mathematicians that are productive in their areas of research have high levels of analytic and practical abilities. Practical abilities manifest in choosing problems that are accessible and publishable. However, creative mathematicians (levels 6 and 7) have higher levels of synthetic abilities in comparison to the level 5 mathematicians in that the papers they publish open up new research vistas for other mathematicians. An example of of such creative work is the aforementioned paper by Hewitt (1948). These high levels of synthetic abilities are perhaps compromised by slightly lower levels of practical abilities.

For example, levels 6/7 mathematicians often leave proofs half-finished or sometimes do not even bother to publish their work.

There are numerous examples in the history of mathematics that reveal such tendencies among highly creative mathematicians. For instance, Srinivasa Ramanujan (1887–1920), the Indian mathematician, had handwritten notebooks filled with numerous theorems without proof that still contributes to fertile directions in the growth of analytic number theory, elliptic functions, infinite series, and continued fractions. David Hilbert's (1900) list of 23 problems presented to mathematicians at the 1900 International Congress in Paris contributed both to the phenomenal growth of mathematics and the particular directions in which it has grown (Rowe & Gray, 2000). The Riemann hypothesis still remains an open problem with profound implications for numerous areas of mathematics. The most recent example is that of Paul Erdös, a contemporary and enigmatic mathematical genius, who was renowned for giving other mathematicians conjectures and/or problems with hints, partial solutions, or no solutions. The mathematicians that finished these problems and wrote up the results were usually gracious to list Erdös as the co-author of their papers. In fact co-authors of Erdös gave themselves a number called "Erdös number 1", and mathematicians that co-authored a paper with "Erdös number 1" mathematicians gave themselves "Erdös number 2" and so on.

Elements of Sternberg's (1997) trirachic view of giftedness are also seen in the model in Figure 4.2. In order to maximize the potential for creativity to manifest in the mathematics classroom, teachers can encourage mathematically creative students to share their synthetic insights on connections between seemingly diverse problems with the other students in the class (Sriraman, 2004a). Historic examples of synthetic thinking in mathematics, which connect seemingly diverse ideas/concepts can be used in the classroom to further illustrate the power and value of such insights. The scholarly, free market and aesthetic principles contain aspects of Sternberg's triachic view of giftedness.

The five principles also encompass notions of polymathy which can foster creativity in general by connecting notions from the arts and sciences to mathematics and vice versa. Common thinking traits of hundreds of polymaths (historical and contemporary) as analyzed by Root-Bernstein (1989, 1996, 2000, 2001, 2003) and many others are: (a) visual geometric thinking and/or thinking in terms of geometric principles; (b) frequent shifts in perspective; (c) thinking in analogies; (d) nepistemological awareness, or an awareness of domain limitations; (e) interest in investigating paradoxes, which often reveals interplay between language, mathematics and science; (f) belief in *Occam's Razor*, or belief that simple ideas are preferable to complicated ones; (g) acknowledgment of serendipity and the role of chance; and (h) the drive to influence the agenda of the times (Sriraman, 2005).

One recent example provided by Root-Bernstein (2003) is the effect of Escher's drawings on a young Roger Penrose, the mathematical physicist, who visited one of Escher's exhibitions in 1954. Stimulated by the seemingly impossible perspectives conveyed by Escher in 2-dimensions, Penrose began creating his own impossible objects such as the famous Penrose "impossible" tribar which shows a 3-dimensional triangle that twists both forwards and backwards in 2-dimensions. Root- Bernstein wrote

Roger Penrose showed his tribar to his father L. S. Penrose, a biologist who dabbled in art... [who] invented the impossible staircase in which stairs appear to spiral both up and down simultaneously... [and] sent Escher a copy... [who] then developed artistic possibilities of the impossible staircase in ways that have since become famous (p. 274).

Another well known consequence of Escher's artistic influence on mathematicians is the investigation of tiling problems (both periodic and aperiodic) popularized by both Roger Penrose and Martin Gardner, which helped cystallographers understand the structure of many metal alloys which are aperiodic (Root-Berstein, 2003).

THE CHANGING NATURE OF MATHEMATICS

Another important aspect of this discussion is the question of the balance between pure and applied mathematics. The literature suggests that the nature of mathematics relevant for today's world has also changed. In spite of the rich and antiquated roots of mathematics, Steen (2001) suggested that mathematician's today should acknowledge the contributions of researchers in external disciplines like biology, physics, finance, information sciences, economics, education, medicine, and so on who successfully use mathematics to create models with far reaching and profound applications in today's world. These interdisciplinary and emergent applications have resulted in the field of mathematics thriving at the dawn of the 21st century.

However, problem solving as it is implemented in the classroom does not contain this interdisciplinary approach and modeling of what is happening in the real world. In the U.S the urgency of preparing today's students adequately for future-oriented fields is increasingly being emphasized at the university level. Steen (2005) writes that "as a science biology depends increasingly on data, algorithms and models; in virtually every respect it is becoming... more mathematical" (p. xi). Both the National Research Council (NRC) and the National Science Foundation (NSF) in the U.S are increasingly funding universities to initiate interdisciplinary doctoral programs between mathematics and the other sciences with the goal of producing scientists who are adept at "mathematizing" reality. Secondary mathematics is usually the gateway for an exposure to both breadth and depth of mathematical topics. However, most traditional mathematics curriculum are still anchored in the traditional treatment of mathematics, as opposed to an interdisciplinary and modeling based approach of mathematics used in the real world (Sriraman, in press). Sheffield, Bennett, Berriozabal, DeArmond, and Wertheimer (1995) lamented that not much had changed in terms of mathematics curriculum as of that point in time and remarked that gifted students of mathematics were the ones that were the most shortchanged and unable to utilize their talents which could be viewed as a societal resource invaluable to maintain leadership in a technologically changing world. Moreover, high school mathematics also serve as the gatekeeper for many areas of advanced study (Kerr, 1997). The traditional treatment of mathematics has little or no emphasis on modeling-based activities, which require team work and communication. Additionally, the traditional mathematics has historically kept gifted girls from pursuing 4 years of high school mathematics. This deficit is difficult to remediate at the undergraduate level and results in the effect of low numbers of students capable of graduate level work in interdisciplinary fields such as mathematical biology and bio-informatics (see Steen, 2005). Any educator with a sense of history should foresee the snowball effect or the cycle of blaming inadequate preparation to high school onto middle school onto the very elementary grades, which suggests we work bottom up. That is, we should initiate and study the modeling of complex systems that occur in real life situations from the very early grades. Lesh, Kaput, and Hamilton (2006, in press) reported that in projects such as Purdue University's Gender Equity in Engineering Project, when students' abilities and achievements were assessed using tasks that were designed to be simulations of real life problem solving situations, the understandings and abilities that emerged as being critical for success included many that are not emphasized in traditional textbooks or tests. Thus, the importance of a broader range of deeper understandings and abilities, and a broader range of students naturally emerged as having extraordinary potential. Surprisingly enough, these students also came from populations, specifically female and minority, that are highly underrepresented in fields that emphasize mathematics, science, and technology, and they were underrepresented because their abilities had been previously unrecognized (Lesh & Sriraman, 2005a, 2005b; Sriraman, 2005).

Thus it may be more fruitful to engage students in model-eliciting activities, which expose them to complex real life systems, as opposed to contrived problem solving. The mathematical conceptual systems arising from such investigations have great potential for being pursued by mathematically gifted students purely in terms of their implications, and because they create axiomatic structures through which theorems can be discovered that are analogous to what a pure mathematician does (Sriraman & Strzelecki, 2004).

CONCLUDING NOTES

In conclusion, the goal of the gifted education community is not simply to ensure that mathematically gifted students fulfill their potential by becoming productive pure and applied mathematicians, but also to ensure that the mathematically creative students among the mathematically gifted do not get overlooked. After all, these may indeed be the very students that have the potential to move the field forward through their atypical/unorthodox methods and insights. The butterfly effect of overlooking one of these potential level 6 (creative mathematician) students in the classroom eventually affects the livelihood of a thousand potential level 5 (productive mathematician) students. A case in point that illustrates the far-reaching ripples of such a butterfly effect is Hilbert's (1900) problems, which sustained both pure and applied mathematics, as well as Ramanujan's notebooks.

REFERENCES

Barnes, M. (2000). Magical moments in mathematics: Insights into the process of coming to know. *For the Learning of Mathematics, 20*(1), 33–13.

Barnett, L. B., & Corazza, L. (1993). Identification of mathematical talent and programmatic efforts to facilitate development of talent. *European Journal for High Ability, 4,* 48–61

Behr, M., & Khoury, H. (1986). Children's inferencing behavior. *Journal for Research in Mathematics Education, 17*(5), 369–381.

Benbow, C. P., Lubinski, D. & Sushy, B. (1996). The impact of SMPY's educational programs from the perspective of the participant. In C. P. Benbow & D. Lubinski (Eds), *Intellectual talent* (pp. 266–300). Baltimore: Johns Hopkins University Press.

Birkhoff, G. D. (1956). Mathematics of aesthetics. In J. R. Newman (Ed.), *The world of mathematics, Vol. 4* (7th ed., pp. 2185–2197). New York: Simon and Schuster.

Birkhoff, G. D.(1969). Mathematics and psychology. *SIAM Review, 11,* 429–469.

Brinkmann, A. (2004). The experience of mathematical beauty. In Contributions to P. C. Clarkson, M. Hannula (Organizers), TSG 24: Students' motivation and attitudes towards mathematics and its study. *Proceedings of the 10th International Congress of Mathematics Education, Copenhagen, Denmark.*

Brower, R. (1999). Dangerous minds: Eminently creative people who spent time in jail. *Creativity Research Journal, 12*(1), 3–14.

Burton, L. (1999a). The practices of mathematicians: What do they tell us about coming to know mathematics? *Educational Studies in Mathematics, 37*(2), 121–143.

Burton, L. (1999b). Why is intuition so important to mathematics but missing from mathematics education? *For the Learning of Mathematics, 19*(3), 27–32.

Chambers, J. A. (1964). Relating personality and biographical factors to scientific creativity. *Psychological Monographs, 78*(7), 584.

Chang, L. L. (1985). Who are the mathematically gifted elementary school children? *Roeper Review, 8*(2), 76–79.

Craft, A. (2003). The limits to creativity in education: Dilemmas for the educator. *British Journal of Educational Studies, 51*(2), 113–127.

Craft, A. (2002). *Creativity in the early years: A lifewide foundation.* London: Continuum.

Cramond, B. (1994). Attention-deficit hyperactivity disorder and creativity—What is the connection? *Journal of Creative Behavior, 28,* 193–210.

Creme, P. (2003). Why can't we allow students to be more creative? *Teaching in Higher Education, 8*(2), 273–277.

Csikszentmihalyi, M. (1988). Society, culture, and person: A systems view of creativity. In R. J. Sternberg (Ed.), *The nature of creativity: Contemporary psychological perspectives* (pp. 325–339). Cambridge University Press.

Csikszentmihalyi, M. (2000). Implications of a systems perspective for the study of creativity. In R. J. Sternberg (Ed.), *Handbook of creativity* (pp. 313–338). Cambridge University Press.

Davis, G. A. (1997). Identifying creative students and measuring creativity. In N. Colangelo & G. A. Davis (Eds.), *Handbook of gifted education* (pp. 269–281). Boston: Allyn Bacon.

Davis, P. J., & Hersh, R. (1981). *The mathematical experience.* New York: Houghton Mifflin.

Davydov, V. V (1988). The concept of theoretical generalization and problems of educational psychology. *Studies in Soviet Thought, 36,* 169–202.

Davydov, V. V. (1990). Type of generalization in instruction: Logical and psychological problems in the structuring of school curricula. In J. Kilpatrick (Ed.), *Soviet studies in mathematics education, Vol. 2,* Reston, VA: National Council of Teachers of Mathematics.

Diezmann, C., & Watters, J. (2003). The importance of challenging tasks for mathematically gifted students. *Gifted and Talented International, 17*(2), 76–84.

Dreyfus, T., & Eisenberg, T. (1986). On the Aesthetics of Mathematical Thought. *For the Learning of Mathematics, 6*(1), 2–10.

Einstein, A., & Inheld, L. (1938). *The evolution of physics.* New York: Simon and Schuster.

English, L. D. (in press). Problem posing in the elementary curriculum. In F. Lester, & R. Charles (Eds.), *Teaching mathematics through problem solving.* Reston, Virginia: National Council of Teachers of Mathematics.

Ervynck, G. (1991). Mathematical creativity. In D. Tall (Ed.). *Advanced mathematical thinking* (pp. 42–53). Kluwer Academic Publishers.

Frensch, P., & Sternberg, R. (1992). *Complex problem solving: Principles and mechanisms.* Mahwah, NJ: Erlbaum.

Goldberg, A., & Suppes, P. (1972). A computer assisted instruction program for exercises on finding axioms. *Educational Studies in Mathematics, 4,* 429–449.

Greenes, C. (1981). Identifying the gifted student in mathematics. *Arithmetic Teacher, 28*(6), 14–17.

Gruber, H. E. (1989). The evolving systems approach to creative work. In D. B. Wallce & H. E. Gruber, *Creative people at work: Twelve cognitive case studies.* Oxford: Oxford University Press.

Gruber, H. E. (1981). *Darwin on man*. Chicago: University of Chicago Press.

Gruber, H. E., & Wallace, D. B. (2000). The case study method and evolving systems approach for understanding unique creative people at work. In R. J. Sternberg (Ed.), *Handbook of creativity* (pp. 93–115). Cambridge University Press.

Hadamard, J. W. (1945). *Essay on the psychology of invention in the mathematical field*. Princeton, NJ: Princeton University Press.

Hardy, G. H. (1940). *A mathematician's apology*. London.

Heid, M. K. (1983). Characteristics and special needs of the gifted student in mathematics. *The Mathematics Teacher, 76,* 221–226.

Hershkowitz, R. (1989). Visualization in geometry—Two sides of the coin. *Focus on Learning Problems in Mathematics, 11,* 61–76.

Hewitt, E. (1948). Rings of real-valued continuous functions. *Transactions of the American Mathematical Society, 64,* 45–99.

Hilbert, D. (1900). Mathematische Probleme: Vortrag, gehalten auf dem internationalen Mathematiker-Congress zu Paris 1900. *Gött. Nachr.* 253–297.

Jackson, L. (2002). *Freaks, geeks and Asperger Syndrome: A user guide to adolescence*. London: Jessica Kingsley.

James, I. (2003). Austism in mathematicians. *The Mathematical Intelligencer, 25*(4), 62–65.

Kajander, A. (1990) Measuring mathematical aptitude in exploratory computer environments. *Roeper Review, 12*(4), 254–256.

Kanevsky, I. S. (1990). Pursuing qualitative differences in the flexible use of a problem solving strategy by young children. *Journal for the Education of the Gifted, 13,* 115–140.

Kerr, B. A. (1997). Developing talents in girls and young women. In N. Colangelo & G. A. Davis (Eds.), *Handbook of gifted education* (2nd ed., pp. 483–497). Boston: Allyn & Bacon.

Kieren, T., & Pirie, S. (1991). Recursion and the mathematical experience. In L. P. Steffe (Ed.) *Epistemological foundations of mathematical experience* (pp. 78–102). New York: Springer-Verlag.

Kiesswetter, K. (1985). Die förderung von mathematisch besonders begabten und interessierten Schülern—ein bislang vernachlässigtes sonderpädogogisches problem. *Der mathematische und naturwissenschaftliche Unterricht, 38,* 300–306.

Kiesswetter, K. (1992). Mathematische Begabung. Über die Komplexität der Phänomene und die Unzulänglichkeiten von Punktbewertungen. *Mathematik-Unterricht, 38,* 5–18.

Krutetskii, V. A. (1976). *The psychology of mathematical abilities in school children*. (J. Teller, trans. & J. Kilpatrick & I. Wirszup, Eds.). Chicago: University of Chicago Press.

Kuhn, T. S. (1962). *The structure of scientific revolutions*. Chicago: University of Chicago Press.

Lesh, R., Kaput, J., & Hamilton, E. (Eds.) (2006, in press), *Foundations for the Future: The Need for New Mathematical Understandings & Abilities in the 21st Century*. Hillsdale, NJ: Lawrence Erlbaum Associates.

Lesh, R., & Sriraman, B. (2005a). John Dewey Revisited- Pragmatism and the models-modeling perspective on mathematical learning. In A. Beckmann, C. Michelsen & B. Sriraman (Eds). *Proceedings of the 1st International Symposium on*

Mathematics and its Connections to the Arts and Sciences. (pp. 32–51). May 18–21, 2005, University of Schwaebisch Gmuend: Germany.Franzbecker Verlag,

Lesh, R., & Sriraman, B. (2005b). Mathematics education as a design science. *International Reviews on Mathematical Education (Zentralblatt für Didaktik der Mathematik)*, *37*(6), 490–505.

Lester, F. K., & Kehle, P. E. (2003). From problem solving to modeling: The evolution of thinking about research on complex mathematical activity. In R. Lesh & H. Doerr (Eds.) *Beyond constructivism: Models and modeling perspectives on mathematics problem solving, learning and teaching* (pp. 501–518). Mahwah, NJ: Erlbaum.

Marshak, D. (2003). No child left behind: A foolish race into the past. *Phi Delta Kappan*, *85*(3) 229–231.

Massé, L., & Gagné, F. (2002). Gifts and talents as sources of envy in high school settings. *Gifted Child Quarterly*. *46*(1), 15–29.

Minsky, M. (1985). *The society of mind.* New York: Simon & Schuster Inc.

National Council of Teachers of Mathematics (1989). *Curriculum and standards for school mathematics*, Reston, VA: Author.

Nickerson, R. S. (2000). Enhancing creativity. In R. J. Sternberg (Ed.), *Handbook of creativity* (pp. 392–430). Cambridge University Press.

Plucker, J., & Beghetto, R. A. (2004). Why creativity is domain general, why it looks domain specific, and why the distinction does not matter. In R. J. Sternberg, E. L. Grigorenko & J. L. Singer (Eds.), *Creativity: From potential to realization* (pp. 153–168). Washington DC: American Psychological Association.

Policastro, E., & Gardner, H. (2000). From case studies to robust generalizations: An approach to the study of creativity. In R. J. Sternberg (Ed.), *Handbook of creativity* (pp. 213–225). Cambridge University Press.

Poincaré, H. (1948). *Science and method.* New York: Dover.

Polya, G. (1945). *How to solve it.* Princeton, NJ: Princeton University Press.

Polya, G. (1954). *Mathematics and plausible reasoning: Induction and analogy in mathematics* (Vol. II). Princeton University Press.

Presmeg, N. C. (1986). Visualization and mathematical giftedness. *Educational Studies in Mathematics, 17*, 297–311.

Renzulli, J. S. (1978). What makes giftedness? Reexamining a definition. *Phi Delta Kappan, 60*, 180–184, 261.

Renzulli, J. S. (1986). The three-ring conception of giftedness: A developmental model for creative productivity. In R. J. Sternberg & J. E. Davidson (Eds.), *Conceptions of giftedness* (pp. 332–357). New York: Cambridge University Press.

Ripple, R. E. (1989). Ordinary creativity, *Contemporary Educational Psychology, 14*, 189–202.

Root-Bernstein, R. S. (1989). *Discovering.* Cambridge, MA: Harvard University Press.

Root-Bernstein, R. S. (1996). The sciences and arts share a common creative aesthetic. In A. I. Tauber (Ed.), The *elusive synthesis: Aesthetics and science* (pp. 49–82). Netherlands: Kluwer.

Root-Bernstein, R. S. (2000). Art advances science. *Nature, 407,* 134.

Root-Bernstein, R. S. (2001). Music, science, and creativity. *Leonardo, 34,* 63–68.

Root-Bernstein, R. S. (2003). The art of innovation: Polymaths and the universality of the creative process. In L. Shavanina (Ed.), *International handbook of innovation*, (pp. 267–278), Amsterdam: Elsevier.

Rowe, D., & Gray, J. (2000). *The Hilbert Challenge.* Oxford University Press.

Schoenfeld, A. (1985).*Mathematical problem solving.* Lawrence Erlbaum & Associates

Schoenfeld, A. H. (1993). Learning to think mathematically: Problem solving, meta-cognition, and sense making in mathematics. In D. Grouws (Ed.) *Handbook of research on mathematics teaching and learning* (pp. 334–370). New York: Mc-Millan.

Shapiro, S. I. (1965). A study of pupil's individual characteristics in processing mathematical information. *Voprosy Psikhologii*, No. 2.

Shaw, M. P. (1994). Affective components of scientific creativity. In M. P. Shaw & M. A. Runco (Eds.), *Creativity and affect* (pp. 3–43), Norwood, NJ: Ablex.

Sheffield, L. J., Bennett, J., Berriozabal, M., DeArmond, M., & Wertheimer, R. (1995). *Report of the task force on the mathematically promising.* Reston, VA: National Council of Teachers of Mathematics.

Silver, E. A. (Ed.) (1985). *Teaching and learning mathematical problem solving: Multiple research perspectives.* Hillsdale, NJ: Erlbaum.

Smith, J. M. (1966). *Setting conditions for creative teaching in the elementary school.* Boston: Allyn and Bacon.

Sowell, T. (2001). *The Einstein Syndrome.* New York: Basic Books.

Sriraman, B (in press). Implications of research on mathematics gifted education for the secondary curriculum. To appear in C. Callahan & J. Plucker (Editors) *What the research says: Encyclopedia on research in gifted education.* Prufrock Press.

Sriraman, B. (2002). How do mathematically gifted students abstract and generalize mathematical concepts. *NAGC 2002 Research Briefs, 16*, 83–87.

Sriraman, B. (2003). Mathematical giftedness, problem solving, and the ability to formulate generalizations. *The Journal of Secondary Gifted Education. 14*(3), 151–165.

Sriraman, B. (2004a). Reflective abstraction, uniframes and the formulation of generalizations. *The Journal of Mathematical Behavior, 23*(2), 205–222.

Sriraman, B. (2004b). Discovering a mathematical principle: The case of Matt. *Mathematics in School, 33*(2), 25–31.

Sriraman, B. (2004c). The characteristics of mathematical creativity. *The Mathematics Educator, 14*(1), 19–34.

Sriraman, B. (2004d). Gifted ninth graders' notions of proof. Investigating parallels in approaches of mathematically gifted students and professional mathematicians. *Journal for the Education of the Gifted, 27*(4), 267–292.

Sriraman, B. (2005). Philosophy as a bridge between mathematics arts and the sciences. In A. Beckmann, C. Michelsen, & B. Sriraman et al (Eds.), *Proceedings of the 1st International Symposium on Mathematics and its Connections to the Arts and Sciences* (pp. 7–31). May 18–21, 2005, University of Schwaebisch Gmuend, Germany: Franzbecker Verlag.

Sriraman, B., & English, L. (2004). Combinatorial mathematics: Research into practice. *The Mathematics Teacher, 98*(3), 182–191

Sriraman, B., & Strzelecki, P. (2004). Playing with powers. *The International Journal for Technology in Mathematics Education, 11*(1), 29–34.

Steen, L. A. (2001). Revolution by stealth. In D. A. Holton (Ed). *The teaching and learning of mathematics at university level* (pp. 303–312). Kluwer Academic Publishers: Dodrecht.

Steen, L. A. (2005). *Math & Bio 2010: Linking undergraduate disciplines.* Mathematical Association of America.

Sternberg, R. J. (1997). A triarchic view of giftedness: Theory and Practice. In N. Colangelo & G. A. Davis (Eds.), *Handbook of gifted education* (pp. 43–53). Boston: Allyn Bacon.

Sternberg, R. J., & Lubart, T. I. (1996). Investing in creativity. *American Psychologist, 51,* 677–688.

Sternberg, R. J., & Lubart, T. I. (2000). The concept of creativity: prospects and paradigms. In R. J. Sternberg (Ed.)., *Handbook of creativity* (pp. 93–115). Cambridge University Press.

Suppes, P., & Binford, F. (1965). Experimental teaching of mathematical logic in the elementary school. *The Arithmetic Teacher, 12,* 187–195.

Torrance, E. P. (1981). Non-test ways of identifying the creatively gifted. In J. C. Gowan, J. Khatena, & E. P. Torrance (Eds.), *Creativity: Its educational implications* (2nd ed., pp. 165–170). Dubuque, IA: Kendall/Hunt.

Torrance, E. P. (1974). Torrance tests of creative thinking: Norms-technical manual. Lexington, MA: Ginn.

Usiskin, Z. (2000). The Development into the Mathematically Talented. *Journal of Secondary Gifted Education, 11*(3), 152–162.

Vitale, B. (1989). Elusive recursion: A trip in a recursive land. *New Ideas in Psychology, 7*(3), 253–276.

Vygotsky, L. (1962). *Thought and language.* Cambridge, MA: MIT Press.

Vygotsky, L. (1978). *Mind in society: The development of higher psychological processes.* Cambridge, MA: Harvard University Press.

Wallas, G. (1926). *The art of thought.* New York: Harcourt Brace.

Wason, P. C., & Johnson-Laird, P. N. (1972). *Psychology of reasoning.* Cambridge, MA: Harvard University Press.

Wertheimer, M. (1945). *Productive Thinking.* New York: Harper.

Weisberg, R.W. (1993). *Creativity: Beyond the myth of genius.* New York: Freeman.

Yakimanskaya, I. S. (1970). Individual differences in solving geometry problems on proof. In J. Kilpatrick & I. Wirszup (Eds.). *Soviet studies in the psychology of learning and teaching mathematics (Vol. 4),* Stanford: School Mathematics Study Group.

Ypma, E.G. (1968). Predictions of the industrial creativity of research scientists from biographical information. *Dissertation Abstracts International, 30,* 5731B–5732B.

ACKNOWLEDGMENT

Reprint of Sriraman, B. (2005). Are Giftedness and Creativity Synonyms in Mathematics? *Journal of Secondary Gifted Education, 17*(1), 20–36. Reprinted with permission from Prufrock Press. ©2005 Bharath Sriraman.

NOTES

1. These scores place the students approximately in the 95–99 percentile band.
2. The Fields Medals was established by John Charles Fields (1863–1932) and is the equivalent of the Nobel Prize for the field of mathematics. These medals are awarded every four years to mathematicians under 40 years of age, at the International Congress of Mathematics.
3. Behr and Khoury (1986) found that the inferencing behavior of younger school children were analogous to those found by Wason and Johnson-Laird (1972).
4. The Riemann hypothesis states that the zeros of Riemann's zeta function all have a real part of one half. Conjectured by Riemann in 1859 and since then has neither been proved nor disproved. This is currently the most outstanding unsolved problem in mathematics.
5. The Bieberbach conjecture is easily understood by undergraduate students with some exposure to complex analysis because of the elementary nature of its statement. A univalent function f transforms a point in the unit disk into the point represented by the complex number $f(z)$ given by an infinite series $f(z) = z + a_2 z^2 + a_3 z^3 + a_4 z^4 + \ldots$ where the coefficients $a_2, a_3, a_4, \ldots$ are fixed complex numbers, which specify f. In 1916 Bieberbach conjectured that no matter which such f we consider $|a_n| \le n$. Loius de Branges proved this in 1985.

CHAPTER 5

DOES MATHEMATICS GIFTED EDUCATION NEED A WORKING PHILOSOPHY OF CREATIVITY?

Viktor Freiman
Université de Moncton, New Brunswick

Bharath Sriraman
The University of Montana

ABSTRACT

In this paper, we present existing views and approaches to creativity with emphasis on their links with mathematics. Educational and social views of creativity in general, and mathematical creativity in particular is discussed. While institutional and societal indifference to the needs of the mathematically gifted is a well known phenomena, recent research studies reveal that even less attention is paid to the development and nurturing of creativity in the mathematically gifted. We will discuss the need of such particular attention from the mathematician's, psychologists' and educators' points of view. We will conclude our discussion with recommendations for the kind of learning and teaching environment for mathematically gifted students necessary to

Creativity, Giftedness, and Talent Development in Mathematics, pages 113–132

113

stimulate and nurture their creativity as well as to benefit the needs of other students in that setting.

INTRODUCTION

The constructs of mathematical giftedness and mathematical creativity are inter-connected, with creativity implicitly implying giftedness (Sriraman, 2005). However studying a mathematician's or a student's creativity is a very difficult enterprise because most traditional operationalized instruments fail to capture extra cognitive traits such as beliefs, aesthetics, intuitions, intellectual values, self imposed subjective norms, spontaneity, perseverance standards, and chance which contribute towards astonishing acts and products of creative endeavors (Shavinina & Ferrari, 2004).

A BRIEF OVERVIEW OF THE LITERATURE

Sternberg's Classification of Approaches to the Study of Creativity

The Handbook of Creativity (Sternberg, 2000), which contains a comprehensive review of all research available in the field of creativity suggests that most the approaches used in the study of creativity can be subsumed under six categories, which are: mystical, pragmatic, psychodynamic, psychometric, cognitive and social-personality. Each of these approaches was reviewed in chapter 1 of this monograph in addition to confluence views of creativity (see chapter 1, Sriraman, 2008)

View of Creativity Expressed by Social Institutions and the Community

The literature summarized in chapter 1 does not take into full consideration cross-cultural differences in views of what constitutes mathematical creativity. Cultural and social aspects play a significant role in what the community, in general, and the school system, in particular, considers as "creativity" and how they deal with it. Numerous studies (Crammond, 1994; Davis, 1997; Smith, 1966; Torrance, 1981) indicate that the behavioral traits of creative individuals very often go against the grain of acceptable behavior in the institutionalized school setting. For instance, negative behavioral traits such as indifference to class rules, display of boredom, cynicism or hyperactivity usually result in disciplinary measures as opposed to appropriate affective interventions. In the case of gifted students who 'conform' to the norm these students are often prone to hide their intellectual capacity for social reasons, and identify their academic talent as being a source of envy (Massé & Gagné, 2002). History is peppered with numerous examples of creative individuals described as "deviants' by the status quo. The stifling of creativity at the K–12 levels is often collectively rationalized under the guise of doing what is supposedly good for the majority of the students, or invoking the oft-misused term "equity," or appealing to curricular plans and school achievement goals etc. etc. The recent passing of the No Child Left Behind Act (NCLB) in the United States under the guise of 'equity' has brought to the forefront the debate of what is to be done with creative and gifted students in the classroom. Recently Marshak (2003) wrote that the NCLB call for accountability based on standardized testing for the traditional skills of reading, writing and arithmetic valued in the industrial societal setting is a giant step *backwards* to the 1940s. Based on recent reports released by the U.S. Department of Labor, Marshak (2003) further states that besides the three "traditional" R's, numerous additional skills such as problem solving and creative thinking skills are necessary for success in the global societal setting of 21st century. Even at the tertiary levels there have been criticisms about the excessive amount of structure imposed on disciplines by academics (Creme, 2003, p. 273). In summary, a significant proportion of the literature indicates that creativity is viewed as a fringe commodity, tolerated and nurtured by some teachers and typically not encouraged. However, such a viewpoint may simply be a function of culture and location.

Australian researchers report a similar situation with education in general and education of mathematically gifted in particular. In their analysis of the current situation, Diezmann and Waters (2002) state that despite all the discourses about the need of a "clever country" and increasing value of the role of creative individuals, the situation doesn't seem to be better 100 years after it was mentioned that the country looks like "the paradise of mediocrity and the grave of genius." In fact, the authors refer to the Australian Senate report on education of gifted children that acknowledges that focusing on minimum standards could have a disastrous effect on satisfying the special needs of the gifted who are already affected by "underachievement, boredom, frustration, and psychological distress, (and)…negative attitudes and mistaken beliefs." This situation is particularly dire in the case of mathematics where gifted children are affected in multiple ways by the generally negative attitude of others towards mathematics. This attitude identifies gifted children as a "marked" group or "deviant" population.

Negative community attitudes, towards the gifted, reaches its extremes in derisive labelling of such children as "little Einsteins" or "nerds." In such a general and pervasive anti-intellectual atmosphere that seems unfavourable to mathematically gifted students these children need particular support and resilience. Based on this background of the existing situation in school systems, which seems to be in accord with the public sentiment one can ask what can be done for gifted children to help them to become more creative. In the next section, we present three particular points of view: psychological, mathematical and educational.

THE NEED FOR NURTURING AND SUPPORT OF CREATIVITY IN THE MATHEMATICALLY GIFTED

Psychologist's Point of View of Mathematical Abilities Related to Creativity

In his longitudinal study on mathematical ability, Krutetskii (1976) argues that a successful mathematical activity requires particular combination of personal traits. He argues that having superior mathematical abilities does not necessarily allow the gifted individual to reach higher mathematical summits. His findings are based on the study of a group of very talented pupils of different ages, biographies of renowned mathematicians, research literature and survey distributed among practicing teachers and professional mathematicians. The paper by Karp in this monograph also addresses Krututeskii's work from a contemporary viewpoint relevant for teacher education.

First, Krutetskii refers to the work of Myasishev (Мясищев) who stated that one can not become a creative mathematician without enjoying mathematical work. Joy of mathematics boosts a desire to search, mobilizes hard working habits and dynamics. Second, the personality of teachers plays an important role. Sometimes, a very able student may not show any special interest in the subject or high results. But if teacher succeeds to discover her hidden talent and boost student's interest this student can become very successful as seen in the biographies of Lobatchevskii, Ostrogradskii, Luzin and others.

The next factor revealed by Krutetskii (1976) is related to the emotional nature of mathematical activity saying that all gifted students in his study showed very high levels of emotions when they succeed to solve a difficult problem or made a mathematical discovery. Also, Krutetskii (1976) emphasizes the importance of aesthetical values of mathematical work. Citing Revesh, who said that "A mathematician creates because the beauty of mental constructions brings him joy," Krutetskii mentions that for his students a good solution made them happy in the same way as a nice combination in chess. Their entire look showed an enjoyment: their eyes were lightened;

they were rubbing their hands and they called each other to share especially nice solutions.

Hard work is another quality considered by Krutetskii (1976) as crucial for creativity in mathematics. In fact, as it was mentioned by Lavrentijev quoted by Krutetskii, the main condition of mathematical creativity is the ability to work hard constantly, over a long period of time. Often, it takes months, years and even decades for mathematician to reach her goal looking constantly for a way to solve a problem, trying to find a better one among 1000 others. These features have been observed also in gifted students from Krutetsii's experimental group.

To complete his list of personal characteristics of gifted and creative mathematicians Krutetskii (1976) brings another three factors. In fact, in order to be creative, a mathematician must also be innovative and have courage not only to create something new but at the same time destroy old established knowledge. At the same time, a creative mathematician has to be critical of her work. Especially regarding gifted students, we need to be careful of not giving them too much credit but rather educating them on the value of critical evaluation of her work and her habits. Krutetskii (1976) finishes with the importance of not focusing only on mathematics but to develop a more harmonious (well-rounded) personality in order to be creative.

Mathematician's Point of View of the Role of Creativity in Mathematical Discoveries

Miller (1997) analyses Poincaré's conception of "sensible intuition" as a process driven by the ability to perceive the whole of the argument at a glance that allows for a selection and assembling of the appropriate combination of mathematical facts. This occurs using the "unconscious" rules of aesthetics and intuition going beyond pure logic to get some sense of the steps in a mathematical proof without access to visual imagery. According to Poincaré, a "special aesthetic sensibility" helps mathematicians to filter few combinations that are "harmonious" and "beautiful" making intuition as an ingredient of creativity. Due to this creative component, mathematicians work on their invention building a network of thought connecting elements from widely separated domains, the process which is unconscious, subtle and delicate that must be "felt rather than formulated" leading to unexpected combination of mathematical facts and to scientific invention.

What conditions are to be met in order to help gifted students to be more creative? First one can learn from reflections of renowned mathematicians when they share their moments of discovery. Gnedenko (1991) analyses different situations in which he was making his own discoveries: one when he posed a new task to himself (related to the Taylor's series) and found a solution in a different way. Another example is related to Lusin's work on trigonometric series. Gendenko was reading an article written by another mathematician and found another solution missed by the author. The third example is related to his learning experience from Hinchin's seminars on special topics of his own research interest when Hinchin succeeded to attract his students with fresh open new problems. Gnedenko lists several conditions to foster mathematical creativity in students. These conditions are (a) creation of special research atmosphere as a source of intellectual inspiration, (b) being a part of a team that works on real novel complex problems, (c) presence of teachers whose approach is based on patient and passionate guidance with some important hints and advice and not telling students the solutions. Finally, intrinsic motivation is also very important as it helps to mobilise inner forces and passion of intellectual hard work on certain within long period of time (ability to focus and to concentrate). Seeing hard work with several iterations, sometimes failing to get immediate result are also necessary conditions for creative mathematical work. Gnedenko argues that the solution to a problem may come as a sudden "aha" idea that might be seen as easy and effortless inner "vision." He refers to examples from writing of Poincaré, Gauss and Hadamard.

Gnedenko gives examples of such sudden insights that happened in his life in very different situations not related to any mathematical activity: during teaching, shopping, traveling or even during night. One example is striking: after 3 days of useless search for a solution to one problem, he told to his teacher (Hinchin) about his doubt of the correctness of a featured conjecture. Coming home, he was very excited, he could not eat, talk to people—his brain was occupied with the problem. While thinking of a problem, he fell asleep and when he woke up, the ready-to-write proof was already in his head. Analyzing the cause of this "sudden" insight Gnedenko argued that even though he was asleep the brain continued to work on an unconscious level. But this "work" has to be prepared by previous process (often seemingly inefficient).

In this sudden aha effect, Gnedenko sees a parallelism between mathematical creation and poetry: both mathematicians and poets are trying to catch up with a "burn-up bird" (жар птица), invisible and hardly achievable and both reach it often quite suddenly but this is preceded by long thinking and searching. Finally Gnedenko mentioned that creative mathematical work has to be related to discovery of something completely novel. He cites the Russian poet V. Mayakovskii who said that the person who found for the first time that two plus two gives four by say putting two matches with two other matches is a great mathematician.

Educators' Point of View on Development Creativity in Gifted Students

According to Gnedenko (1991) mathematical giftedness (математичес-кая одаренность) is not as rare in humans as one might think. But this personal trait of creativity can appear in different ways in different people. One person could be interested in generalizing and more profound examination of already obtained results. Another person shows an ability to find new objects for study and to look for new methods in order to discover their unknown properties. The third type of person can focus on logical development of theories demonstrating extraordinary sense of awareness of logical fallacies and flaws. A fourth group of gifted individuals would be attracted to hidden links between seemingly unrelated branches of mathematics. The fifth would study historical processes of the growth of mathematical knowledge. The sixth would focus on the study of philosophical aspects of mathematics. The seventh would search for ingenious solutions of practical problems and look for new applications of mathematics. Finally, someone could be extremely creative in the popularisation of science and in teaching.

Thus, Gnedenko (1991) relates giftedness directly with creativity. He recognizes that everybody can possess a certain degree of creativity but educational systems and background (school, family, etc.) might become an obstacle to the gifted person if the surrounding system rejects novelty and discourages efforts to look at new aspects of the problem or to go beyond known facts. . The teaching approach can also lead to some obstacles for fostering of mathematical giftedness if the teacher doesn't pay enough attention to gifted students who might loose their interest in going further in their learning of mathematics.

The history of Soviet mathematics provides with a striking example of a coexistence of two different approaches to mathematics education, one embedded into the general lay public educational system implementing the blueprint based on the European concepts of the late 19th century, and the other one focusing mainly on gifted children and having flourished starting from 1950s onwards. The latter took the form of a complex network of activities including "mathematics clubs for advanced children" (Russian "кружки" (*kruzhki*), lit. "circles" or "rings," usually affiliated with schools and universities but some were also home-based), Olympiads, team mathematics competitions, (*mat-boi*, literally "mathematical fight"), extra-curricular winter or summer schools for gifted children, publication of magazines on physics and mathematics for children (the most famous being the *Kvant*, lit "Quantum"), among others (Freiman & Volkov, 2004).

All these activities were free for all participating children and were based solely on the enthusiasm of mathematics teachers or university professors.

This process led to the creation of a system of formation of a "mathematical elite"[1] in the former USSR focused first and foremost on "gifted children," which was in a sharp contrast with the "egalitarian" regular state-run schools targeting "average student" and thus neglecting the needs of all those above average level. This situation was not new to the USSR's educational system but was rooted in the former tsarist Russia' regular school system which usually did not pay much attention to gifted children. Only a few gifted children such as young A. Kolmogorov were able to benefit from a unique extra-curriculum pedagogical environment allowing them to enjoy the beauty of mathematical discovery. He attended a small private school organized by his grand-mother at home for a small group of students of various ages in which the teachers used the most recent pedagogical innovations. A. Kolmogorov (1988) witnesses that at the age of 5 or 6 he was pleased with his discovery of the regularity of the sum of odd numbers: $1 = 1^2$, $1 + 3 = 2^2$, $1 + 3 + 5 = 3^2$, etc. The report of the Kolmogorv's "mathematical discovery" was published in the school magazine.

An interesting episode in Kolmogorov's story is related to his further schooling at a private gymnasium organized by "radically oriented intellectuals."[2] A mixed school (for girls and boys) with the curriculum of "boys' gymnaslum" (i.e., a college) gave the students an opportunity to study according to their own interests and levels (Kolmogorov could for example, take a math course from one grade higher). At the same time, students felt responsible to study hard and to get best results for the tests and state examinations. It is not surprising that a school of this kind was in contradiction with "regular" state schools and thus was under constant threat of closing by the officials.

After the revolution of 1917, the new Soviet government closed all private schools and established a completely new school system with new curriculum. The system was aimed at offering a basic education to the entire population and, at the same time, making the education more practice-oriented. As result of the implementation of these ideas, "the mathematics program lost much of its theoretical content. Pupils studied mathematical "recipes" applicable in specific practical situations, often without consideration of their theoretical bases."[3] It remains unclear what the situation was of gifted students in those years, yet the sources point at the lack of knowledge displayed by those who graduated from the schools based on such "innovative" approaches as the so-called "brigade-project" organization (Vogeli, 1968 with reference to Bradis, 1954, p. 38).

As a reaction to this gloomy situation, the government declared these methods as "errors" and ordered to make necessary corrections in the school curriculum in the early '30s. Thus, "pre-Revolutionary mathematics texts were resurrected, revised, and made official standard" (Vogeli, 1968).

Resuming our analysis, we can state that keeping an explicitly egalitarian approach to education, the Soviet education system at one moment (namely, in earlier 1930s) began spending much more money and effort to identify promising individuals and to provide them with opportunities to develop their talents (Blazer, 1989).

As the officials struggled to meet the needs of growing economy and to maintain the access to schools for everybody, numerous initiatives came from prominent mathematicians and scientists. One of the striking examples was the first mathematical Olympiad for schoolchildren organized in biggest the Soviet cities: Leningrad, Tbilisi, Moscow in 1934–35.[4] The Olympiads helped to build traditions that went far beyond the officially stated goals (such as a high quality education). Instead, as their participants recall, they became actual festivals of mathematicians of all generations, schoolchildren, university students, school teachers, young high-school professors, and prominent scientists.

The Olympiad problems were not oriented on a mere application of school knowledge but required a capacity to find original ways of thinking, ability to reason logically in a non-standard situation. The Olympiad was usually followed by a lecture with analysis of typical mistakes and by individual meetings of the participants with the members of the jury. The Olympiad was not the only way to work informally with young talents but also a way to motivate school children to learn mathematics in a more systematic way by participating in "mathematics circles," attending public lectures given by outstanding mathematicians and by self-study of mathematical books.

Looking at the social context of this phenomenon, we can consider it as a personal mission of mathematicians contributing to the society in order to promote and popularize mathematics and emphasize the value of creative mathematical work, to search and to support young talents and give them the best of their knowledge. A mathematical community was created beyond regular educational system, and the explicitly stated goal of this community was to maintain the highest level of mathematics and to promote attractiveness of mathematical activity among the population and to support and encourage everyone with talent and interest in mathematics.

In their enrichment philosophy, Diezmann and Watters (2000) stress the need of creating opportunity for gifted children to become creative individuals. For these authors, to be creative, the individual needs an intellectual autonomy, an expertise, and a culture supportive of unconventional thought. In their study, Diezmann and Watters (2000) experimented with a special extra school science program in order to maximise the growth of creativity in gifted children basing on the development of the autonomy and domain based expertise in a social context of recognition and support. The autonomy enabled individuals to deal with novelty and generate creative products in both, evolutionary and revolutionary types of the progress.

In the same order of ideas, the authors stress the importance of good thinking dependent of the problem solving context which requires either a strict application of heuristics, uncritical acceptance of information along with a disregard for contradictory evidence or it promotes non sequential process with cycles of interpretation, intuitions, and testing ideas which are typical characteristics of ill-structured problems. Finally, development of creativity relies on a social context in which gifted individuals get recognition from family and teachers of their abilities and care of their development. In addition, Diezmann and Watters (2000) put emphasis on a collaborative and socially interactive learning environment as a necessary social context.

TOWARDS A MORE INCLUSIVE SCHOOL SYSTEM

Bringing Creativity in Teachers Didactic Inventory for Gifted Children

So far, we could see that in several educational systems opportunity for gifted and creative individuals were created outside or beyond the regular systems. In this section, we will analyze several in-classroom options that should be used by teachers to promote the development and nurturing of creativity in a more inclusive way.

Cline (1999) stresses the need of translating research on the creative individual, creative processes and contexts that promote creative behaviours into classroom practices providing students with opportunities to develop and demonstrate their creative abilities. According to Yastrebov (2005), the inductive nature of mathematical creativity is not being taken into account by teachers. The good understanding of dualistic nature of mathematics needs to be developed in young learners. Each mathematical fact is created by individuals. The existence of mathematics is impossible outside of specific social institution called scientific community. The scientific community approves every mathematical invention. The newly discovered mathematical fact has to be exchanged within the community being critical examined by experts in the related domain.

Guerra, Gimenez, and Servat (2005) point at familiarity, divergence and reinvention as necessary components for teacher's pedagogical knowledge. More precisely, familiarity means proposing potentially creative tasks through identification of unconventional proposals, recognizing variety of models and meanings in different contexts, openness to a variety of answers and surprising results. The divergence component enhances open debates and divergent questioning in different contexts and situations. The reinvention strategy allows choosing adequate didactic sequences to develop invention from the contexts, thinking of imaginative, real and innovative tasks, rediscovering previously learned mathematical knowledge in a new way.

Such role of teachers as promoters of creative mathematical work is crucial for developing of gifted students. According to Karp (2007, this issue), special attention has to be brought to prospective teachers' education which should enhance their didactic knowledge with examples of the ways in which mathematically gifted students construct their knowledge and use it for their further creative activity.

Many authors point at the necessity of creating more challenging mathematical classroom environments which would be suitable for all students including the gifted ones.

Creativity and Thinking As Components of Nurturing Learning-Teaching Environment for Gifted Children

Cline (1999) points at four thinking abilities to be developed which are associated with creativity, namely divergent thinking; fluency; flexibility; originality and elaboration which are fundamental elements of Guilford's (1967) definition of divergent thinking as the generation of information from given information putting emphasis on variety and quantity of output from the same source also involving transfer . In the following paragraphs, we will relate these definition to the mathematics educators' perspective on mathematical thinking.

Many authors point at the *ability to recognise patterns and to see relationships* as a *key element* in mathematical thinking. Fisher argues (1990) that since mathematics is a highly structured network of ideas, to think mathematically is to *form connections* in this network: the task of a teacher then is to help children to see the structure inherent in mathematics, not just rules and facts learned in isolation. He states that in encouraging children to think mathematically we need to engage all aspects of a child's intelligence. There are different ways of processing mathematics according to the following scheme (Figure 5.1).

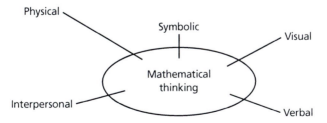

Figure 5.1   Different ways of mathematical thinking.

We see that, according to this model, mathematical problems can be modelled or represented in a variety of ways:

- *Verbally*: through inner speech and talking things through, using linguistic intelligence, putting planning procedures and process into words, making sense and meaning for oneself.
- *Inter-personally*: learning through collaboration observing others, working together to achieve a shared goal, exchanging and comparing ideas, asking questions, discussing problems.
- *Physically*: using physical objects in performing mathematical tasks, working with practical apparatus, equipment and mathematical tools, modelling a problem or process, having hands-on experience, using bodily-kinesthetic skills, practical applications into the physical world
- *Visually*: putting processes into pictorial form, making drawings or diagrams visualising patterns and shapes in the mind's eye, thinking in spatial terms, graphical communication, geometric designs, manipulating mental images.
- *Symbolically*: using written words and abstract symbols to interpret, record and work on mathematical problems, using different recording systems, logically exact languages, translating into mathematical codes.

According to Baroody (1987), genuine learning also involves the ability in *changing of thinking patterns*. In fact, organized differently, insights can provide fresh and more powerful perspectives thus changing *how* a child thinks about something. More specifically, making a connection can change the way knowledge is organised. The child who does not know the basic subtraction combinations uses fingers to calculate differences. For example facing the following series of subtractios: $2 - 1 = \_$, $4 - 2 = \_$, $6 - 3 = \_$, $8 - 4 = \_$, $10 - 5 = \_$, the child laboriously calculates each answer. Suddenly the child may have an insight: These subtraction combinations are the inverses of well-known additions of doubles ($1 + 1 = 2$, $2 + 2 = 4$, $3 + 3 = 6$, $4 + 4 = 8$, and $5 + 5 = 10$). Thus she would produce a new relationship between subtraction combinations and the familiar addition facts which allows her to see subtraction in a different light. Given a problem like $5 - 3 = \_$, the child now thinks to herself: "Three plus what makes five? Oh, yeah, two." Her new perspective now enables her to solve subtraction combinations efficiently without laborious calculation. Mathematical development, then, entails qualitative changes in thinking as well as quantitative changes in the amount of information stored. Essential to the development of understanding are changes in thinking patterns (Baroody, 1987, p. 11).

According to Schrag (1988), mental activity is *purposeful* thinking only if it is experienced as *directed* to a problem or task one has set oneself. This, admittedly normative, conception is meant to include cases in which we may suddenly see a solution without any awareness of "wrestling" with a problem. But even in such cases, an idea does not appear as a *solution* unless it is experienced in relation to some difficulty one has been worrying about. Thus thinking is evoked in *situations* where one is not quite sure how to go on. Schrag calls such situations *problems*.

Referring to works of Polya (1957) and Schoenfeld (1979), Ernst (1998) points at two types of thinking activities that might affect problem-solving behaviour: cognitive and metacognitive. *Cognitive* activities include using and applying facts, skills, concepts and all forms of mathematical knowledge. They also include applying general and topic specific mathematical strategies, and carrying out problem-solving plans. *Meta-cognitive* activities, involve planning, monitoring progress, making effort calculations (e.g., "Is this approach too hard or too slow?"), decision making, checking work, choosing strategies, and so on. Metacognition (literally: "above cognition") is about the management of thinking. Sierpinska, Ninadozie, and Octac (2002) characterize mathematical thinking as "a good balance between *theoretical* and *practical* thinking." In their study of relationships between theoretical thinking and high achievement in linear algebra, Sierpinska et al. (2002) assume that theoretical thinking "is not a continuation but a reversal of the practical thought." (p. 11). They view practical thinking as an "epistemological obstacle" that cannot and should not be avoided. However, they claim that teaching abstract mathematical concepts that puts too much emphasis on the "concrete" experience based on so-called "geometric" or "numerical" approaches might leave students with representations irrelevant from the point of view of the concepts and lead them to contradictions. (Ibid, p. 19).

Defining theoretical thinking as reflective, systemic (definitional, proof-based and hypothetical) and analytic (linguistic sensitive and meta-linguistic sensitive), these authors argue for the necessity of theoretical thinking in understanding linear algebra as following:

- The undergraduate learner of linear algebra must be even more theoretically inclined than the inventors of the theory.
- Meanings of concepts must be sought in their relations with other concepts.
- The learner must engage in proving activity and therefore use systemic approaches to meanings and validity.
- The learner has to accept that his or her ontological questions will remain unanswered.
- The learner must engage in hypothetical thinking.
- The learner must become mathematically "multilingual." (Ibid, 33–35)

If we project these ideas on the elementary school level, we could see that today's tendencies (discussed in previous sections) do not favor the education of a "theoretical thinker" although our practice shows that mathematically gifted students, even at an early age hold certain epistemological views about mathematics that are close to the theoretical thinking.

When we think about the interpretation of these aspects of thinking in terms of giftedness, then we may suppose that a mathematically gifted child who is a high achiever would probably demonstrate a balanced ability to think mathematically (theoretically and practically). A mathematically able child who is not a high achiever would be rather more "theoretically inclined." The question is whether a mathematically able child can be only "practical"? Another question is to what point we can identify a child as a theoretical thinker?

Another question arises: what kind of classroom situations would enhance the fostering the development of mathematical thinking in young children?

In order to *foster* the development of mathematical thinking, Baroody (1993) stresses a use of a problem-solving approach which focuses on the processes of mathematical inquiry: problem solving, reasoning, and communicating. It is a teacher-guided approach in which a student plays an active role.

Ernst (1998) makes a comparative analysis of different teaching approaches related to the mathematical thinking. It shows the didactical transition of mathematical process which "progresses from the application of facts, skills and concepts, to a limited repertoire of problem-solving strategies including generalisation and the induction of pattern, to the full range of problem-solving strategies, and finally adding problem-posing processes as well"(p. 132) happens when the classroom teaching becomes more open and challenging.

Fischbein (1990) defines a teacher's task as to "create an environment that would require a mathematical attitude, mathematical concepts, and mathematical solutions." He argues that when facing a challenging task, children might not be able to find solutions spontaneously. They might get engaged in a constructional process combining various conditions. They then have to produce a method to work on the problem systematically. Fischbein sees this aspect of finding a method, an algorithm used consciously as fundamental for the development of mathematical reasoning. According to Fischbein (1990), the question is whether the teacher should wait until children find the method by themselves without any help. In his opinion:

> Formal reasoning doesn't develop spontaneously as a main way of thinking. This conclusion doesn't imply that the teacher should simply offer the solution. What the teacher should do is to direct the student's efforts to a solu-

tion by asking adequate questions. The student builds the answers as a reaction to a certain environment. This environment should be programmed as a problematic one in order to inspire student's solution endeavors. (Fischbein, 1990, p. 8)

These theoretical guidelines cohere with Driscoll's (1999) remarks that through
- Consistent modeling of algebraic thinking.
- Giving well timed pointers to students that help them shift or expand their thinking, or that help them to pay attention to what is important.
- Making it a habit to ask a variety of questions aimed at helping students organize their thinking and respond to algebraic prompts.

Teachers would promote those habits of mind that are specific to the algebraic thinking and should be developed in children as follows:

- Reversibility as a capacity not only to use a process to get to the goal, but also to understand the process well enough to work backward from the answer to the starting point.
- Building rules as a capacity to *recognize patterns* and *organize data.*
- *Abstracting from calculations* as a capacity to think about computations independently of particular numbers that are used. (Driscoll, 1999, p. 3)

CONCLUDING REMARKS

Drawing from the aforementioned theoretical considerations, we now move towards more practically oriented questions such as what are mathematical activities that would help to foster the emergence of a creative thinking in mathematically gifted elementary school students allowing them to progress in the mixed-ability classroom? Numerous studies point at mathematically rich tasks as an engine of such fostering. Peressini and Knuth (2000) mention that mathematically rich tasks are those that fit following criteria:

- Encourage a range of solution approaches,
- Address significant mathematical concepts,
- Require students to justify their explanations,
- Are open ended

The use of such tasks requires a rethinking of the role of both the teacher and the student. Burton (1984) stresses that the teacher's role shifts from that of providing information, to question-asker and resource-provider.

The teacher would challenge pupils to justify or falsify arguments and to reflect on what has been done. The tone of the teacher's interventions is also important. It has to emphasise enquiry rather than instruction. Sriraman's (2004) discussion of mathematical creativity indicates that many of the characteristics of mathematical creativity described by mathematicians as invaluable aspects of their craft such as the freedom to choose and pursue problems in an academic setting, the freedom of movement required during work, the awareness of the distinction between learning versus creating, the aesthetic appeal of mathematics and the affective urgency/drive to solve problems with tremendous real world implications, might be difficult to simulate in a traditional classroom setting. However there are basic principles which are implementable by classroom teachers. Five overarching principles that emerged from Sriraman's (2005) synthesis and analysis of the literature as significantly enhancing mathematical creativity are labeled as (a) The Gestalt principle, (b) The Aesthetic principle, (c) The Free Market principle, (d) The Scholarly principle, and (e) The Uncertainty principle.

As for the student, we underline that along with sustaining interest, motivation and success in problem-solving, the following aspects are also to be mentioned:

- Choosing and using representations that would enhance the ability to model a problem.
- Resolving instead of solving which helps to develop an attitude to build a network of new questions, new resolutions, and further questions out of an initial problem following a spiral development instead of a linear one (problem—solution).
- Communicating using different tools of communication.

Our study of challenging situation approach within an inclusive mathematics classroom (Freiman, 2006) demonstrates that in such situations, gifted children can go further, go beyond situations, ask new questions, initiate their own investigations, and be more creative in their mathematical work. At the same time, such classroom environments turn to be nurturing for all students helping them to become more creative learners.

Many researchers in mathematics education stress an importance of specific didactical environments to nurture creative thinking in young learners. Meissner (2005) concentrates his study on three aspects:

- Individual and social components, like motivation, curiosity, self-confidence, flexibility, engagement, humor, imagination, happiness, acceptance of oneself and others, satisfaction, success, ...

- Profound discussions as well as intuitive or spontaneous "*challenging problems*" that are fascinating, interesting, exciting, thrilling, important, provoking, . . .
- The students must be able to identify themselves with the problem and its possible solution(s), developing *important abilities* to explore and to structure a problem, to invent own or to modify given techniques, to listen and argue, to define goals, to cooperate in teams, . . .

Taking these aspects into consideration would help children to become active, to discover and to experiment, to enjoy and to have fun, to guess and to test, to laugh at their own mistakes. The recent ICMI study 16 on challenging mathematics led by P. Taylor and E. Barbeau, http://www.amt.canberra.edu.au/icmis16dd.html accentuates the need for further research on providing students with rich and challenging mathematical problem solving experiences in order to develop and nurture their creativity.

REFERENCES

Baroody, A. (1987). *Children's mathematical thinking: a developmental framework for preschool, primary and special education teachers.* Columbia: Teachers College, Columbia University.

Baroody, A. (1993). Problem *solving, reasoning, and communication, K–8: helping children think mathematically.* Macmillan Publishing Company.

Birkhoff, G. (1969). Mathematics and psychology. *SIAM Review,* 11, 429–469.

Blazer, H. D. (1989). *Soviet Science on the edge of reform,* Boulder: Westview Press.

Bradis, V. M. (1954). *Methodology of teaching mathematics in the secondary school,* Utchpedgiz, In Russian.

Burton, L. (1984). *Thinking Things Through.* Oxford: Basil Blackwell Limited.

Cline (1999). *Giftedness has many faces: multiple talents and abilities in the classroom.* The Foundation of Concepts in Education, Inc., 193 pp.

Cramond, B. (1994). Attention-deficit hyperactivity disorder and creativity- What is the connection? *Journal of Creative Behavior,* 28, 193–210.

Csikszentmihalyi, M. (1988). Society, culture, and person: A systems view of creativity. In R. J. Sternberg (Ed.), *The nature of creativity: Contemporary psychological perspectives* (pp. 325–339). Cambridge University Press.

Csikszentmihalyi, M. (2000). Implications of a systems perspective for the study of creativity. In R. J. Sternberg (Ed.), *Handbook of creativity* (pp. 313–338). Cambridge University Press.

Davis, G. A. (1997). Identifying creative students and measuring creativity. In N. Colangelo & G. A. Davis (Eds.), *Handbook of Gifted Education* (pp. 269–281). Boston: Allyn Bacon.

Davis, P. J., & Hersh, R. (1981). *The mathematical experience.* New York: Houghton Mifflin.

Diezmann, C., & Watters, J. (2000). An enrichment philosophy and strategy for empowerment young gifted children to become autonomous learners. *Gifted and Talented International 15*(1), 6–18.

Diezmann, C., & Watters, J. (2002). Summing up the Education of Mathematically Gifted Students. *Proceedings of the 25th Annual Conference of the Mathematics Education Research Group of Australasia*, pp. 219–226.

Driscool, M. (1999) *Fostering Algebraic Thinking: A Guide for Teachers, Grades 6–10.* Portsmouth, NH: Heinemann.

Ernst, P.(1998). Recent development in mathematical thinking. In R. Burden, & M. Williams (Eds.), *Thinking through the curriculum* (pp. 113–134). London, New York: Routledge.

Fischbein, E. (1990). Introduction. In: P. Nesher, & J. Kilpatrick (Eds.), *Mathematics and cognition: A research synthesis by the International Group for the Psychology of Mathematics Education.* Cambridge: University Press.

Fisher, R.(1990). *Teaching children to think,* Oxford: Basil Blackwell.

Freiman, V. (2006). Problems to discover and to boost mathematical talent in early grades: A challenging situations approach. *The Montana Mathematics Enthusiast, 3*(1), 51–75

Freiman, V., & Volkov, A. (2004). Early Mathematical Giftedness and its Social Context: The Cases of Imperial China and Soviet Russia. *Journal of Korea Society of Mathematical Education Series D: Research in Mathematical Education,* 8(3),157 173.

Gallian, J. A. (1994). *Contemporary Abstract Algebra.* Lexington, MA: D.C. Heath and Co.

Gnedenko B. V. (1991) *Introduction in specialization: mathematics* (Введение в специальность: математика), Nauka, 235 pages. In Russian

Gruber, H. E., & Wallace, D. B. (2000). The case study method and evolving systems approach for understanding unique creative people at work. In R. J. Sternberg (Ed.), *Handbook of creativity* (pp. 93–115). Cambridge University Press.

Guerra, Gimenez, & Servat (2005). *Detecting Traits of Creativity Potential In Mathematical Tasks With Prospective Primary Teachers, ICMI – EARCOME-3* (East Asia Regional Conference on Mathematics Education), Shanghai, China, August, 7–12, 2005, http://www.math.ecnu.edu.cn/earcome3/

Hadamard, J.W. (1945). *Essay on the psychology of invention in the mathematical field.* Princeton University Press.

Karp, A. (2007). Knowledge as a Manifestation of talent: Creating Opportunities for the Gifted. *Mediterranean Journal for Research in Mathematics Education,* this issue.

Kolmogorov (1959). *On the profession of a mathematician.* Moscow State University Press (in Russian).

Krutetskii V. A.(1976). *The psychology of mathematical abilities in school children.* Chicago: The University of Chicago Press.

Lester, F. K. (1985). Methodological considerations in research on mathematical problem solving. In E. A. Silver (Ed). *Teaching and learning mathematical problem solving. Multiple research perspectives* (pp. 41–70). Hillsdale, NJ: Erlbaum.

Marshak, D. (2003). No child left behind: A foolish race into the past. *Phi Delta Kappan,* 85(3) 229–231.

Massé, L., & Gagné, F. (2002). Gifts and talents as sources of envy in high school settings. *Gifted Child Quarterly.* 46(1), 15–29.

Meissner, H. (2005) Creativity and mathematics education. *Paper presented at The 3rd East Asia Regional Conference on Mathematics Education* http://www.math.ecnu. edu.cn/earcome3/sym1/sym104.pdf

Miller, A. (1997) Cultures of Creativity: Mathematics and Physics. *Diogenes, 45,* 53–75

Minsky, M. (1985). *The society of mind.* New York: Simon & Schuster Inc.

Peressini D., & Knuth, E. (2000). The role of tasks in developing communities of mathematical inquiry. *Teaching Children Mathematics,* 391–396.

Poincaré, H. (1948). *Science and method.* Dover: New York.

Polya, G. (1954). *Mathematics and plausible reasoning: Induction and analogy in mathematics* (Vol. II). Princeton, NJ: Princeton University Press.

Polya, G. (1957). *How to solve it.* Princeton, NJ: Princeton University Press.

Ridge, L., & Renzulli, J. (1981). Teaching Mathematics to the Talented and Gifted. In V. Glennon (Ed.) *The mathematics education of exceptional children and youth, an interdisciplinary approach* (pp. 191–266). NCTM.

Sheffield, L. (2003). *Extending the challenge in mathematics.* TAGT & Corwin Press, 150 pp.

Schoenfeld, A. H. (1979). Explicit heuristic training as a variable in problem-solving performance. *Journal for Research in Mathematics Education, 10,* 173–187.

Schoenfeld, A. H.(1985a). *Mathematical problem solving.* New York: Academic Press.

Schrag, F. (1988). *Thinking in School and Society.* New York: Routledge.

Shavinina, L.V., & Ferrari, M. (2004). *Beyond knowledge: Extracognitive aspects of developing high ability.* Mahwah, NJ: Erlbaum.

Sierpinska, A., Nnadozie, A., & Octaç, A.(2002). A study of relationships between theoretical thinking and high achievement in linear algebra. Montreal: Concordia University.

Smith, J. M. (1966). *Setting conditions for creative teaching in the elementary school.* Boston: Allyn and Bacon.

Sriraman, B. (2004). The characteristics of mathematical creativity. *The Mathematics Educator, 14*(1), 19–34.

Sriraman, B. (2005). Are Mathematical Giftedness and Mathematical Creativity Synonyms? A theoretical analysis of constructs. *Journal of Secondary Gifted Education, 17*(1),20–36.

Sriraman, B. (2008). The characteristics of mathematical creativity. (this issue).

Sternberg, R. J. (2000). *Handbook of creativity.* Cambridge University Press.

Sternberg, R. J., & Lubart, T. I. (1996). Investing in creativity. *American Psychologist, 51,* 677–688.

Sternberg, R. J., & Lubart, T.I. (2000). The concept of creativity: prospects and paradigms. In R. J. Sternberg (Ed.), *Handbook of creativity* (pp. 93–115). Cambridge University Press.

Torrance, E. P. (1974). *Torrance tests of creative thinking: Norms-technical manual.* Lexington, MA: Ginn.

Torrance, E. P. (1981). Non-test ways of identifying the creatively gifted. In J.C. Gowan, J. Khatena, & E.P. Torrance (Eds.), *Creativity: Its educational implications* (2nd ed.) (pp. 165–170). Dubuque, IA: Kendall/Hunt.

Vogeli, B. (1968). *Soviet Secondary Schools for the Mathematically Talented.* NCTM.

Weisberg, R.W. (1993). *Creativity: Beyond the myth of genius.* New York: Freeman.

Wertheimer, M. (1945). *Productive Thinking.* New York: Harper.

Yastrebov, A.V. (2005) *Dualisticheskiye svoystva mathematiki I ih otragenije b processe prepodavanija.* Pedagogicheskii vestnik.

ACKNOWLEDGMENT

Reprint of Freiman, V., & Sriraman, B. (2007). Does mathematics gifted education need a philosophy of creativity?. In B. Sriraman (Guest Editor), *Mediterranean Journal for Research in Mathematics Education*, 6(1 & 2), 23–46. Reprinted with permission from The Cyprus Mathematical Society. ©2007 Viktor Freiman & Bharath Sriraman

NOTES

1. A similar process took place in other disciplines, in particular, physics, but in the present paper we will focus on mathematics education.
2. "Радикально настроенная интеллигенция," that is, a group of highly educated persons with radical progressive views and highest demands concerning the quality of the education given to their children.
3. Vogeli mentions that "the 1921 syllabus emphasized the value of creative activity in teaching of mathematics, the need to broaden pupils' mathematical background, and the desirability of relating mathematics to life," Vogeli (1968, p. 4).
4. *The Encyclopedia of Young Mathematician* (Энциклопедический словарь юного математика, Moscow: Pedagogika, 1985, in Russian, p. 187) mentions the famous mathematicians B. N. Delone and P. S. Alexandrov as the intiators and organizers of these competitions.

CHAPTER 6

ENABLING MORE STUDENTS TO ACHIEVE MATHEMATICAL SUCCESS

A Case Study of Sarah

Sylvia Bulgar
Rider University

ABSTRACT

As the United States struggles to compete with other industrialized nations in the teaching and learning of mathematics, attention needs to be focused upon one of our greatest human resources, those who would become distinguished in mathematics. Therefore, it is imperative that we provide all students with optimal prospects for demonstrating superior performance so that they can receive appropriate training throughout their educational careers. In this paper, the author shares some of the mathematical experiences of one child, Sarah, her granddaughter, who shows characteristics of a high level of mathematical achievement. These indicators initially appeared when Sarah was quite young.

Creativity, Giftedness, and Talent Development in Mathematics, pages 133–154

INTRODUCTION AND THEORETICAL FRAMEWORK

In order to comply with the United States No Child Left Behind Act of 2001, states have been working diligently to create measurable educational standards and to select or create instruments of measurement for these standards. Federal subsidies for education are linked to compliance with this legislation, which focuses on having all children meet basic standards of proficiency. Meeting the standards is determined by expanded testing at all grade levels. This has resulted, in many cases, in teachers redefining their teaching goals so that all of their students, regardless of abilities, will be proficient on standardized tests (Schorr & Bulgar, 2003). As of the 2000–2001 academic year, all 50 of the United States and the District of Columbia have a large-scale testing program in place (Education Week on the Web, 2002). The trend toward more extensive testing for standardized proficiency, then again, goes beyond the United States (Abrantes, 2001; Firestone & Mayrowetz, 2000; Keitel, & Kilpatrick, 1998 as cited in Abrantes, 2001; Niss, 1996).

At the same time, the report *"Road Map for National Security: Imperative for Change"* (2001), also known as the Hart-Rudman Commission Report, informed the United States of the critical need to develop a new generation of citizens who are skillful in the areas of science, technology, engineering and mathematics (known as the STEM pipeline). Thus, the STEMEd Caucus was formed by Congressman Vern Ehlers (R-MI) and Congressman Mark Udall (D-CO) to help supply the United States with the intellectual capital needed to sustain and further develop a knowledge-based economy. Consequently, it becomes essential to provide all children, from an early age, with the opportunities that will enhance their performance in mathematics so that their education will support their continued mathematical growth and development.

In spite of the focus on having students develop their mathematical skills, international and national assessments of educational progress in mathematics, such as TIMSS (Schmidt, McKnight & Raizen, 1996) and NAEP (NCES, 2003), document the fact that not enough students in the United States have built an understanding of the ideas or knowledge upon which the procedures, algorithms or problems they are asked to work on, are based. Therefore, it is not surprising that there is an educational focus on identifying and helping struggling students, often through procedural remediation, in order to achieve satisfactory results on tests, without appropriate consideration of the need to build knowledge that can strengthen understanding. Additionally, because of the financial incentives tied to having all students reach a level of proficiency, the gifted population (Goodkin, 2005) and those who have already reached proficiency face potential neglect because of the misconception that they will naturally proceed on their own. These children also need to have their learning goals met, which are not consistent with mere proficiency. Their abilities also need to be honed, not only for their personal development, but also for the good of the nation.

The National Council of Teachers of Mathematics (NCTM) has published the Principles and Standards for School Mathematics (2000) which defines the mathematics that *all* children should learn. Traditionally, children with exceptional ability in mathematics have been identified through high scores on standardized tests. Stanley (1976) indicates the inadequacy of many such tests because tests based upon age or grade level attendance do not have enough "ceiling" for very advanced students. An appropriate test to identify accelerated aptitude would give students an instrument to show their abilities to reason and communicate mathematically at very high levels. To this end, the *Test of Mathematical Abilities for Gifted Students* (TOMAGS) was developed in consideration of both The National Council of Teachers of Mathematics Standards and the characteristics of mathematically gifted students that several researchers have identified. An extensively compiled list of characteristics of children who are gifted in mathematics follows because it is pertinent to this study.

Characteristics of mathematically gifted students (Ryser, & Johnsen, 1998):

- Having the ability to both recognize and spontaneously formulate problems, questions and problem-solving steps (Greenes, 1981; O'Conner & Hermelin, 1979; Scruggs & Mastropieri, 1984; Scruggs, Mastropieri, Monson & Jorgensen, 1985; Sternberg & Powell, 1983).
- Being able to distinguish between relevant and irrelevant information in novel problem-solving tasks (Marr & Sternberg, 1986).
- Being able to see mathematical patterns and relationships (Cruikshank & Sheffield, 1992; Miller, 1990).
- Having more creative strategies for solving problems (Devall, 1983; Dover & Shore, 1991; Miller, 1990; Shore, 1986).
- Being more flexible in handling and organizing data (Cruikshank & Sheffield, 1992; Devall, 1983; Dover & Shore, 1991; Greenes, 1981; Miller, 1990; Shore, 1986).
- Being able to offer original interpretations (Greenes, 1981; Sriraman, 2003).
- Being able to transfer ideas generalized among mathematical situations (Cruikshank & Sheffield, 1002; Greenes, 1981; Miller, 1990; Sriraman, 2005).
- Being intensely curious about numeral information (Cruikshank & Sheffield, 1992; Miller, 1990).
- Being able to learn and understand mathematical ideas quickly (Dover & Shore, 1991; Miller, 1990). Being reflective and taking

longer when solving complex problems or those with several solutions (Davidson & Sternberg, 1984; Miechenbaum, 1980; Sternberg, 1982; Wong, 1982; Woodrum, 1975).

- Being persistent in find the solution to a problem (Ashley, 1973; House, 1987; Sriraman, 2003,2004).
- Displaying speed and flexibility in the use of metacognitive knowledge (Dover & Shore 1991).

It is not expected that any child who is determined to be gifted in mathematics would possess all of these characteristics. However, upon identifying these characteristics, it is possible to use them as benchmarks in identifying mathematically advanced children. These attributes can also provide a window into the work of children who may not necessarily be considered to be gifted, since demonstrating these traits while performing mathematical tasks brings to the forefront the high level of achievement possible for many children. In reviewing the attributes, it is obvious that the focus is on reasoning and problem solving rather than on skillfulness with computation, which is consistent with the NCTM Standards (NCTM 2000) for all children.

To broaden the study of mathematical achievement, some of the existing literature related to affect and mathematical performance is examined. *Affect* (as indicated here) refers to the complex structures of emotional responses, feelings, motivation, attitudes, beliefs and values, as these interact with cognition. Mathematically powerful affect is distinct from mathematically positive affect. Mathematically powerful affect is the ability to perform mathematics *powerfully*, an evident characteristic of mathematical precociousness. It involves *both* positive feelings about mathematics (e.g. curiosity, enjoyment, elation in relation to mathematical insight, pride, satisfaction) and ambivalent or negative feelings (e.g. annoyance, impatience, frustration, anxiety, nervousness, fear). When one possesses mathematically powerful affect, the negative feelings associated with the mathematics occur in safe contexts, so that the students are able to manage and benefit from these feelings. Thus, frustration with a difficult problem leads to anticipation of learning something new, and increased pride of achievement when the problem is solved (Goldin, in press; Golden 2002; Schorr & Goldin, in preparation). Having *mathematically powerful affect* supports students' ability to be persistent in finding the solution to a problem (Ashley, 1973; House, 1987), being reflective and taking longer when solving complex problems or those with multiple solutions (Davidson & Sternberg, 1984; Miechenbaum, 1980; Sternberg, 1982; Wong, 1982; Woodrum, 1975), which are identified characteristics of mathematically gifted students. Therefore if we can help students to develop mathematically powerful affect, it follows that they will exhibit characteristics of advanced mathematical performance.

Kaufman & Baer (2004) indicate that when students in their study assessed their own creativity, they expressed its level to be consistent across domains except for mathematics. Consistently, Goldin (2004) indicates that mathematics is traditionally perceived as being rational, logical and analytical, thereby precluding those well versed in the domain from having emotional components to their achievement. Looking at the characteristics of mathematically gifted students above, it is clear to see that emotional components, rather than procedural skills, play a role in advanced mathematical performance. In fact, having more creative strategies for solving problems is denoted as being a characteristic of high achievement in mathematics (Devall, 1983; Dover & Shore, 1991; Miller, 1990; Shore, 1986).

The empowerment to construct mathematical ideas comes from what has been termed "assembly" or the creation of representations from cognitive building blocks (Davis, Maher, Martino, 1992). When children take ownership of an idea, new constructions can take place. Children build on previous experiences, creating assimilation paradigms or internal metaphors (Davis & Maher, 1990; R. B. Davis, Maher & Martino, 1992). Many classroom conditions contribute to an environment in which this can take place. For example, children need to be given adequate time to delve deeply into a problem and need to be comfortable taking risks. Davis (1997) advocated an alternative learning environment for the teaching of mathematics, which fosters the connection between the representations in the mind of the teacher and the mind of the student. Further, he stated, "If we invite students to think, we have the obligation to take their ideas seriously." (Davis, 1992, p. 349)

In various research situations, environments have been created where students have been empowered to demonstrate mathematical prowess (for example, Bulgar 2002; Bulgar, 2003b; Bulgar, Schorr, & Maher, 2002; Reynolds, 2005; Steencken 2001). Some of these research environments have been replicated in regular classroom practice, achieving similar favorable outcomes (Bulgar, 2003a; Bulgar, under review; Bulgar, Schorr, & Warner, 2004).

In addition to favorable environments, appropriate tasks need to be selected in order to help more children to achieve high levels of mathematical thinking and learning. Significant in task selection is the issue of maintaining high levels of cognitive demand (Smith, M. S., & Stein, M., 1998; Stein, Smith, Henningsen, & Silver, 2000). In their work, Smith, Stein, Henningsen and Silver categorized tasks into four levels of cognitive demand: memorization, procedures without connections, procedures with connections and doing math.

As well as creating empowering environments and selecting tasks that require a high level of cognitive demand, the nature of teacher questioning plays a large role in helping students to construct mathematical ideas for themselves and to take responsibility for their own learning. Towers (1998) states that in traditional classrooms, the teacher is seen as separated from the student, that teaching and learning have been regarded as discrete entities. Her study examines the role of teacher interventions in the development of students' mathematical understanding. Hiebert and Wearne (1993) categorized the questions that teachers ask into four hierarchal types that provoke developmentally escalating levels of mathematical thinking. Dann, Pantozzi and Steencken (1995) state that appropriate teacher questioning can help to foster justification and student discourse.

In this paper, the author attempts to examine the work of one child, Sarah, her granddaughter, who has been engaged in mathematical activities in informal settings from a very young age, and to gauge the results against the criteria for mathematical giftedness. It is meant to demonstrate that because Sarah was given the opportunities to experience certain types of tasks in very specific environments, she was able to excel in mathematics by demonstrating some of the characteristics associated with giftedness in mathematics and to experience mathematically powerful affect. These experiences began at an early age and it is the hypothesis of the author that if all children were provided with similar opportunities, many would be highly successful in mathematics. This is significant because of the necessity to develop mathematical talent as both a responsibility towards individuals and the development of intellectual assets for the good of the nation.

METHODOLOGY

This paper is based upon both anecdotal data and the authentic written work of one child, Sarah. Notions of how to evoke a similar high level of performance in many children are generalized from careful scrutiny of the data. At the time of the data collection, other than the work done on Ice Cream Sundaes, there was no intention to use them for research purposes. The work was done as a sharing activity between grandmother and grandchild. The analysis of the data involves its comparison to the attributes of mathematical giftedness, defined by several researchers and indicated in the theoretical framework so that these characteristics can serve as points of reference of mathematical ability. A description of the setting and conditions of each vignette is included in its description because of the variety among them.

The subject of this case study is Sarah, a child with strong family roots, who enjoys doing mathematical tasks, especially with her maternal grandmother (the author of this paper), who is a mathematics education professor at a local university. Sarah has had an extremely close relationship with her grandmother since birth that extends well beyond the experience of mathematics. She is also athletic, being an outstanding swimmer, outgoing, having many close friends, plays piano and does well in all school subjects. She attends a private parochial school, where only half of her day is devoted to non-religious studies and lives in a middle class suburban neighborhood.

Sarah's school uses the Scott Foresman Math Textbook Series, which is considered a non-traditional series. In 2005, the series was selected to participate in the United States Department of Education's Evaluation of Early Mathematics Curricula, which is a large-scale study, the purpose of which is to determine the effectiveness of several math programs that show promise for improving mathematical achievement in the early grades. (http://www.pearsoned.com/pr_2006/041906.htm) Sarah does well on class math tests and has scored adequately on standardized math tests, though not as well as she has scored on standardized literacy tests. Her scores on standardized tests in mathematics are not remarkable. There is no gifted program in her school and she does not receive any classroom supplements to challenge and encourage her ability in mathematics. She says math is her favorite subject and she enjoys computation as well as problem-solving activities.

This paper traces some of the mathematical activities in which Sarah has participated over the course of several years. At the time of this writing, Sarah is eight years nine months old.

RESULTS AND DISCUSSION

First Observed Experience Indicating Advanced Mathematical Thinking

Because Sarah spent so much time with her grandmother, a mathematics education professor, she was engaged in counting games and talk about numbers, at a very early age. The first time that Sarah was observed demonstrating precociousness in mathematics was during the summer of 1999, when she was approximately two years eight months old. She was accompanying her grandmother on a one-hour car ride to visit her two female cousins, one of whom is ten months older than she and one of whom is ten months younger than she. Being a very articulate child, she carried on a conversation throughout the trip from the back seat of the car. During the course of that conversation her grandmother indicated that she had brought along a sprinkler that looked like an octopus because it had eight tentacles, each of which sprayed water. A short discussion ensued about what an octopus is and mention was made of how the word "octagon" is related. Suddenly, Sarah became very quiet. Through the rearview mirror, she was observed doing something with her fingers. When asked what she was doing, she stated, "I'm thinking." She was asked to share her thinking when she was finished. She indicated that if the octopus had eight hoses, she, her cousins, Samantha and Olivia, could each have their own hose and they could still have five friends over to each have their own hoses. Sarah spontaneously had created and solved a problem involving subtraction, and a one-to-one correspondence between hoses and children, thereby demonstrating the ability to see mathematical patterns and relationships. These are characteristics of giftedness in mathematics (Davidson & Sternberg, 1984; Greenes, 1981; Miechenbaum, 1980; O'Conner & Hermelin, 1979; Scruggs & Mastropieri, 1984; Scruggs, Mastropieri, Monson & Jorgensen, 1985; Sternberg, 1982; Sternberg & Powell, 1983; Wong, 1982; Woodrum, 1975). This episode also indicates an interest and curiosity about numerical information. She has taken a mathematical fact (the octopus sprinkler has eight hoses) and not only thought that there would be enough for an individual hose for her and each of her cousins, but she has calculated how many hoses would remain. This is quite a mathematical feat for a child so young. The curiosity that guided her to generate, think about and solve the problem is also an indicator of giftedness (Cruikshank & Sheffield, 1992; Miller, 1990).

Caterpillars

In preparation for the teaching of a mathematics methods class, the author perused several mathematical activities to use with her students. One of these was an activity found online (Rotz & Burns, 2002), but was altered slightly for use in her classroom. The problem was intended for use with first graders who are typically six years old. In an effort to see how the alterations would be received, it was administered to Sarah on November 2, 2002, when she was four years eleven months old (See Figure 6.1).

The task design involves a deliberate developmental increase in difficulty as one proceeds. Essentially, there is an effort to elicit a relationship between the number of circles of the caterpillar and its age. This problem can be very revealing about the subject's thinking. It is possible to merely add one circle as one adds one year of age, and in several observed situations that is the way students generally begin. However, when the problem jumps from a seven-year-old caterpillar to a ten-year-old caterpillar, different solution strategies emerge. A child might continue with the arithmetic process of adding one year of age at a time and thereby adding one circle at a time. A more sophisticated method of finding a solution is to look at the problem geometrically, which involves algebraic reasoning. In essence, this means that the student would be looking for a relationship between the number of circles and the number of years of age rather than just adding one to each. According to the late James Kaput, a recognized authority on algebraic reasoning, algebraic reasoning is comprised of representation, generalization, and formalization of patterns and regularity (Kaput, 1998).

Caterpillars

This is a 1 year-old caterpillar. Count the circles. How many circles does it have? _____

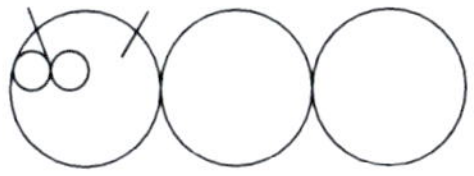

This is a 2 year-old caterpillar. Count the circles. How many circles does it have? _____

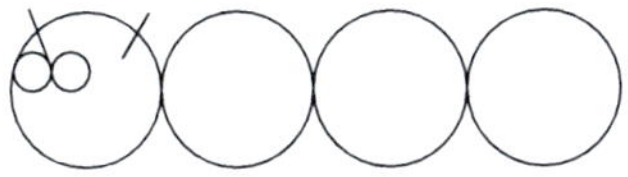

How many circles would a three-year-old caterpillar have? _____
How many circles does a 4-year-old caterpillar have? _____
Draw a 4-year-old caterpillar.

How many circles does a 5-year-old caterpillar have? _____
How many circles does a 6-year-old caterpillar have? _____
How many circles does a 7-year-old caterpillar have? _____
How many circles does a 10-year-old caterpillar have? _____
How do you know? Can you draw a 10-year old caterpillar on the back

Figure 6.1   Caterpillars task.

When facilitating this problem, it is important to observe closely and question carefully to see what method the child uses. When Sarah completed the final portion of the task, getting the answer that the ten-year-old caterpillar has twelve circles, she was asked, "How do you know?" Her answer, recorded on the worksheet was, "because it's ten, eleven, twelve." She does not mention the intermittent ages between seven and ten. She is counting on, or adding two to the age, ten, thus applying algebraic reasoning. Sarah is demonstrating advanced mathematical thinking in several ways here. She is generalizing ideas (Cruikshank & Sheffield, 1002; Greenes, 1981; Miller, 1990); she is recognizing the significance of the relationship between the number of circles and the age of the caterpillar (Marr & Sternberg, 1986) and she is observing and recognizing mathematical patterns and relationships (Cruikshank & Sheffield, 1992; Miller, 1990).

Family Patterns

Sarah often demonstrates an interest in numeracy. For example, she sees patterns in the families of her first cousins, some of which are obvious and some of which are subtler. Soon after her youngest brother was born in

2003, when Sarah was approximately five and one half years old, she began creating "family patterns." In the comparison between the children in her family and her first cousins on her father's side of the family, she recognized that the Girl-Boy-Boy pattern held true for both. However, her maternal first cousins are in the pattern Girl-Girl-Boy. Sarah stated that these cousins did not fit into the same pattern as her own family, but that she could create a pattern that would cross over both families. She arranged the six cousins in order of age and noted that when doing so, the children alternated between families, forming a different pattern, but a pattern nonetheless. That is, if S represents a child from Sarah's family and M represents a child from her mother's sister's family, then Sarah described the following pattern: M-S-M-S-M-S. It would appear that Sarah searches for patterns. Clearly, she demonstrates a propensity for mathematics by displaying identified characteristics of giftedness in mathematics. She appears to be very curious about numerical information and she searches for patterns and relationships (Cruikshank & Sheffield, 1992; Miller, 1990). In addition, her perseverance in finding the second pattern is noteworthy (Ashley, 1973; House, 1987) and she has demonstrated a high level of comfort with the organization of data (Devall, 1983; Dover & Shore, 1991; Miller, 1990; Shore, 1986).

Lollipops and Symbolic Notation

When Sarah was five years five months old and in kindergarten, she accompanied her grandmother to work where they attended an informal meeting. She sat at a table at the far end of the room, outfitted with paper and markers because she said she would color during the meeting. Afterwards, she showed that she was "doing math." She had created simple addition and subtraction problems using symbolic notation. The problems were expressed as fractions with the solutions forming the denominators (See Figure 6.2).

After the meeting, the author, Sarah's grandmother, and another mathematics education professor looked at Sarah's work and were amazed, not merely by its advanced aspects, but also by the fact that Sarah had initiated this work on her own. Some of the numerals are reversed, which is not unusual for this age, but the solutions are correct except for one problem. Sarah had written that $6 - 7 = 0$ just after she wrote that $7 - 6 = 1$. She was questioned about the former of these two problems and asked to explain how she knew this to be true. She indicated that it wasn't true, that $6 - 7$ was not 0. She said that it was less than 0, but she didn't know how to write that. After further questioning she was able to state that it was one less than zero. The notion of numbers existing below zero is extremely advanced for a five year old. She was then told, "Mathematicians have a special name for one less than zero. They call it negative one." Given this language she was able to discern what negative two would mean. Here again, Sarah is demonstrating a high

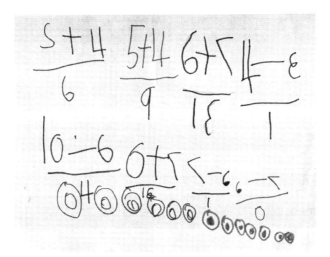

Figure 6.2 Addition and subtraction problems using symbolic notation, which Sarah calls, "Doing Math."

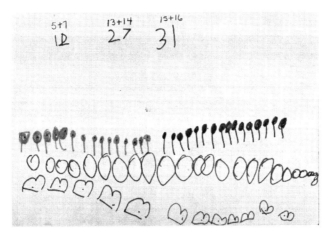

Figure 6.3 Sarah's solutions to the harder problems she requested.

level mathematical achievement by being able to spontaneously formulate problems, questions and problem-solving steps (Greenes, 1981; O'Conner & Hermelin, 1979; Scruggs & Mastropieri, 1984; Scruggs, Mastropieri, Monson & Jorgensen, 1985; Sternberg & Powell, 1983), by being able to offer original interpretations (Greenes, 1981) and by being able to learn and understand mathematical ideas quickly (Dover & Shore, 1991; Miller, 1990).

Upon completion of the originally generated problems and discussion Sarah asked to be given more difficult problems (See Figure 6.3). She was given the expression 5 + 7 to solve. She did this easily and said she wants harder problems, with bigger numbers. She was subsequently given the additional expressions 13 + 14 and 15 + 16. For each of these, she created a representation to help her count. She used hearts of two colors, circles of two colors and lollipops of two colors respectively to represent the addends. Her use of the representations in this way indicates that she understands the meaning of addition as a joining of discrete parts.

Many children learn to add by counting as Sarah demonstrated. However, we see a dot in each of her hearts and in each of her lollipops indicating that when she joined the two addends she did not merely count on to the original number, but had to count all of the items, starting with one. Developmentally, this is common and it enables Sarah to manage much larger numbers.

Discovering Division

When Sarah was six years one month and spending some time with her grandmother, she was rummaging through some workbooks in her grandmother's office. She came across a page that had basic division facts using both of the common symbols, ÷ and ⌐. Never having seen these symbols before, Sarah asked about them and was told they were division symbols. When asked the meaning of division, Sarah was told that it involved finding out how many of one thing can go into another thing and was given a contextual example such as, "If I had six pieces of candy and wanted to share them evenly among three people, how many would each one get?" Sarah wanted to do the page of division facts (See Figure 6.4).

Figure 6.4 Sarah's first experience with division.

Sarah set out to work on the page. She was offered unifix cubes to provide a concrete context for the examples on the page, but she refused them. She solved all of the problems using only the symbolic notation. All but one of her examples were correct. Again, this vignette documents Sarah's mathematical prowess. She has shown her curiosity about numerical information (Cruikshank & Sheffield, 1992; Miller, 1990) and has learned new mathematical ideas quickly (Dover & Shore, 1991; Miller, 1990).

Harry Potter's Chocolate Frogs

In the spring of 2005, the author and her university students attended a workshop on integrating problem solving into a curriculum based upon the use of a textbook. In the workshop the facilitator correlated a problem to a few pages in a third grade textbook. The problem follows here: Harry Potter bought 39 chocolate frogs on the Hogwarts Express. Mrs. Weasley sent him 29 more chocolate frogs for his birthday. How many chocolate frogs did he have all together?[1]

Halfway down the page, there is a horizontal line drawn. Students are told solve the problem above the line. When they are done, they are told to solve it using another method below the line.

Thinking that Sarah would enjoy doing this problem, she was offered a copy of the problem and given the same directions. At the time she was seven years four months and in second grade (See Figure 6.5).

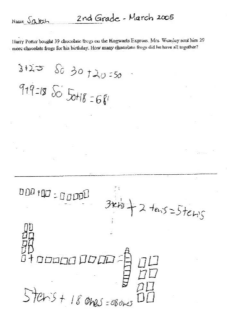

Figure 6.5   Chocolate frogs.

Sarah's first solution involves only symbolic notation. She reasons that since $3 + 2 = 5$, then $30 + 20 = 50$. Though she is only in second grade, she appears to understand the elements of place value and uses this knowledge to solve the problem. In her second solution, below the line, she solves the problem using a representation that she clearly connects to symbolic notation. Thus, Sarah is able to move flexibly between the two solutions and connect them, solving a problem requiring multiple steps (Dover & Shore 1991; Greenes, 1981; O'Conner & Hermelin, 1979; Scruggs & Mastropieri, 1984; Scruggs, Mastropieri, Monson & Jorgensen, 1985; Sternberg & Powell, 1983), which are characteristics of mathematical giftedness.

Ice Cream Sundaes

In order to keep the results of this study current, a task was administered to Sarah as the paper was being written. The task selected is called "Ice Cream Sundaes" and follows directly below:

> You are in an ice cream store where you make your own ice cream dessert. You can choose from any of the following:
>
> - Chocolate ice cream
> - Strawberry ice cream
> - Whipped cream
> - Hot fudge
> - Cherry
>
> How many different ways can you make your ice cream?

This problem has been facilitated many times by the author to elementary school aged children as well as undergraduate university students. Certain patterns emerge in the solution. Younger children tend to create random combinations, often creating duplicates and omitting several. Most undergraduate students eventually find the solution, twenty-four, through various methods such as the use of organization by cases, induction and random listing. It is very common for the undergraduate students to initially omit the cases wherein ice cream is selected alone.

Sarah worked on this task on August 2, 2006 amidst the noise and activity created by the presence of her two brothers and three cousins. Her only concern was that the older one not help her. At the time she was eight years eight months old. She read the problem to herself and was given markers and paper. Her only question was whether or not she was permitted to use letters to stand for the words (abbreviations) so she wouldn't have to keep writing everything out. Throughout her work she used the following abbreviations. She chose to write out the entire word, cherry.

- c.i.c. = chocolate ice cream
- s.i.c. = strawberry ice cream
- H.F. = hot fudge
- W.C. = whipped cream

Sarah began with random listings and very quickly asked if she could start again. When asked why, she indicated that it was too confusing to know what she already wrote. She had apparently recognized the shortcomings of random listing. She then divided the paper by drawing a vertical line down the center and seven horizontal lines across the paper, creating a grid of sixteen boxes. She wrote each combination in another box in the grid, using her abbreviations and created another grid on another piece of paper when her combinations exceeded sixteen. She worked attentively in spite of the noise around her, focusing and persevering until her solution that was complete. When she was finished, she was very confident about her solution of twenty-one ice cream desserts. She had omitted the ones with ice cream alone.

There was no opportunity to interview Sarah about her work until a day later. She was eager to talk about what she had done. She was asked how she came up with her solution and why she was sure it was correct. She stated that she started with a dessert containing everything. She continued by removing one topping, one at a time, resulting in three combinations having two flavors of ice cream and two toppings each. After that she went on to have only one topping on each of three desserts with two flavors of ice cream. She then repeated what she had done with the toppings with chocolate ice cream alone, and finally with strawberry ice cream alone. In essence, Sarah has created a very sophisticated proof by cases, though she has left out one case, the one in which there are no toppings. She used the word, sets, to describe the cases, indicating that her thinking involved a structure for organization. To further organize her work, she used capital letters for the topping abbreviations and lower case letters for the ice cream flavors. Her cases, or sets, are divided by ice cream flavor and her sub-cases are based upon the number of toppings. She is showing her ability to manage and organize data in a highly developed manner, an attribute of mathematical giftedness (Cruikshank & Sheffield, 1992; Devall, 1983; Dover & Shore, 1991; Greenes, 1981; Miller, 1990; Shore, 1986).

Once Sarah explained how she had generated her solution, the following conversation took place between her and the author:

> **A:** Was there something that was included in each of her desserts?
>
> **S:** Ice cream

> **A:** Is there anything else that each dessert had to have?
> **S:** [Big smile] No. I just thought of more desserts.

Sarah then added the ones with no toppings to her grid.

> **A:** What were you thinking that made you smile?
> **S:** That an ice cream dessert could also be plain like when I go to the ice cream store I get plain ice cream.
> **A:** What does this have to do with the problem you worked on?
> **S:** I knew I didn't have all the ice cream desserts [after the question] so I added plain ice cream and strawberry and chocolate ice cream together, with no topping.
> **A:** So how many ice cream desserts did you end up with?
> **S:** Twenty-four.

CONCLUSIONS AND IMPLICATIONS

In this paper, several activities have been described, either anecdotally or with documented supportive data. These activities demonstrate the work of Sarah, the granddaughter of the author of this paper, at various ages. In the analysis of these activities, it becomes apparent that Sarah has, on occasion, demonstrated several attributes of giftedness in mathematics. These attributes can be aligned with the characteristics of giftedness as defined by a wide spectrum of researchers.

Csikszentmihalyi, Rathunde, and Whalen (1997) contend that no matter how gifted someone is, he or she cannot perform as a mathematician unless exposed to experiences involving the mathematical domain. This research points to an extension of that notion by indicating that a child who is not considered to be gifted in mathematics can demonstrate excellence in mathematical performance, under certain conditions. Sarah has had the opportunity to be exposed to various enriching mathematical tasks within safe contexts because her grandmother, who lives very close by and with whom she enjoys spending time, is a mathematics education professor. As she has gotten older, it is possible that Sarah has initiated some of these activities because she is aware of her grandmother's work and wants to forge the connection between them.

If we look at Sarah's work with Chocolate Frogs (See Figure 6.5), we see that she is comfortable finding solutions involving either symbolic notation (numerals) or pictorial representations. The ability to use a variety of representations for the same concept is often associated with the growth of mathematical thinking (Warner & Schorr, 2004). The reader may notice that

Sarah's initial solution involves symbolic notation or numbers, rather than pictures, manipulatives, or other forms of concrete tools. It may be possible to infer that her prior experiences have allowed her to mentally represent the situation, make the connections that would lead her to a more symbolically based solution, one more typical of generalized knowledge. It is only when she is told to find another solution that she retrieves her representational use of pictures.

If Sarah can at times demonstrate some of characteristics of giftedness, how can all children be provided with the opportunity to demonstrate a higher level of achievement in mathematics? Further, how can these opportunities be provided within the milieu of the regular classroom? From the research, one can infer that at least three factors contribute to developing higher levels of mathematical thinking and mathematically powerful affect in many children. They are: the design of appropriate environments, selection of appropriate tasks and suitable teacher interventions. In order to develop mathematically powerful affect, we need to provide safe contexts (Goldin, in press; Golden 2002; Schorr & Goldin, in preparation). Students need to develop empowerment regarding mathematics. These environments have been created in regular classrooms (Bulgar, 2003a; Bulgar, under review). Sarah felt safe while doing the mathematical tasks above, safe enough to take risks and safe enough to restructure her thinking and begin anew when necessary as she did with the Ice Cream Sundaes Problem.

All of the tasks administered to Sarah and discussed above demonstrate examples of activities that make high cognitive demands upon students (Smith, M. S., & Stein, M., 1998; Stein, Smith, Henningsen, & Silver, 2000). The problems were not solved with a mere numerical answer and require justification and proof. The tasks were complex and could not be solved algorithmically. They required Sarah to look at the nature and structure of the problems as well as to monitor her own thinking. These are all characteristics of a task with a high cognitive demand.

The research cited in the theoretical framework is part of a body of knowledge that informs us that in order to facilitate higher levels of mathematical thinking, appropriate interventions need to take place in the classroom. Because such information on teacher interventions exists, there is a need for professional development to enhance teacher understanding of the pedagogy associated with interventions that elicit mathematical empowerment and deeper understanding of mathematical concepts.

Sarah has had the opportunity to experience mathematical problem solving situations that encompassed the factors described above and she has on these occasions demonstrated high levels of mathematical thinking. Additionally, these factors have yielded positive outcomes when integrated into regular teaching practice. Therefore, it is the contention of the author that if knowledge and implementation of the creation of appropriate environments, selection of tasks that evoke high levels of cognitive demand and the use of apt interventions were more widespread, more children would exhibit higher levels of mathematical achievement.

At both the state and national levels we continually hear about the need to encourage citizens to pursue the fields that filter through the STEM pipeline. In order to attract and prepare the most capable members of our society, there is a need to capitalize on the abilities of as many students as possible so that those abilities can be nurtured, honed and enriched throughout their academic careers. This means that we need to provide teachers with the tools necessary to recognize high levels of mathematical achievement, other than exclusively through the use of test scores. By familiarizing school personnel with the characteristics of mathematical giftedness such as those researched-based criteria cited in this paper, these characteristics can be seen as goals for more children. Though this is certainly in the best interest all individuals who would have greater opportunities to excel in mathematics, it is also is crucial for the economy and security of the nation to develop this intellectual capital as a natural resource.

REFERENCES

Abrantes, P. (2001). Revisiting the goals and the nature of mathematics for all in the context of a national curriculum. In M. vandenHeuvel-Panhuizen (Ed.), *Proceedings of the 25th Conference of the International Group for the Psychology of Mathematics Education* (pp. 25–40).

Ashley, R. M. (Ed.). (1973). *Activities for motivating and teaching bright children*. West Nyack, NY: Parker Publishing (as cited in Ryser & Johnsen, 1998).

Bulgar, S. (2002). Through a teacher's lens: Children's constructions of division of fractions. Unpublished doctoral dissertation. Rutgers, The State University of New Jersey, New Brunswick, NJ.

Bulgar, S. (2003a). Children's sense-making of division of fractions. *The Journal of Mathematical Behavior: Special Issue on Fractions, Ratio and Proportional Reasoning, Part B. 22*(3), 319–334.

Bulgar, S. (2003b). Using research to inform practice: Children make sense of division of fractions. In N. A. Pateman, B. J. Dougherty, & J. T. Zilliox (Eds.), *Twenty-seventh Conference of the International Group for the Psychology of Mathematics Education held jointly with the Twenty-fifth Conference of the North American Chapter of the International Group for the Psychology of Mathematics Education: Vol. 2. Navigating Between Theory and Practice* (157–164). Honolulu, HI: CRDG, College of Education, University of Hawai'i.

Bulgar, S. (under review). The development of flexible representations for division of fractions. *Mathematics Teaching and Learning*.

Bulgar, S., Schorr, R. Y. & Maher, C. A. (2002). Teacher's questions and their role in helping students build an understanding of division of fractions. In A. D. Cockburn & E. Nardi (Eds.), *Twenty-sixth Annual Conference of the Inter-*

national Group for the Psychology of Mathematics: Vol. 2. Learning From Learners (pp. 161–167). Norwich, UK: School of Education and Professional Development University of East Anglia.

Bulgar, S., Schorr, R. Y. & Warner, L. B. (2004). Extending and refining models for thinking about division of fractions. *Twenty-sixth Conference of the North American Chapter of the International Group for the Psychology of Mathematics Education: Building Connections Between Communities.* Toronto, Ontario.

Burns, M. (2000). *About teaching mathematics.* Sausalito, CA: Math Solutions Publications.

Cobb, P., Boufi, A., McClain, K., & Whitenack, J. (1997). Reflective discourse and collective reflection. *Journal for Research in Mathematics Education, 28*(3), 258–277.

Cruitshank, D. E., & Sheffield, L. J. (1992). *Teaching and learning elementary and middle school mathematics* (2nd ed.). New York: Macmillan (as cited in Ryser & Johnsen, 1998).

Csikszentmihalyi, M., Rathunde, K., & Whalen, S. (1997). *Talented teenagers: The roots of success and failure.* Cambridge, UK: Cambridge University Press.

Dann, E., Pantozzi, R. S., & Steencken, E. (1995). Unconsciously learning something: A focus on teacher questioning. In *Proceedings of Seventeenth Annual Meeting of the North American Chapter of the International Group for the Psychology of Mathematics Education.* Columbus, Ohio: Ohio State University.

Davidson, J., & Sternberg, R. (1984). The role of insight in intellectual giftedness. *Gifted Child Quarterly, 28,* 58–64 (as cited in Ryser & Johnsen, 1998).

Davis, R. B. (1992). Understanding "understanding." *Journal of Mathematical Behavior, 11,* 225–241.

Davis, R. B. (1997). Alternative learning environments. *Journal of Mathematical Behavior, 16*(2), 87–93.

Davis, R. B., & Maher, C. A., (1990). The nature of mathematics: What do we do when we do mathematics. In R. B. Davis, C. A. Maher, & N. Noddings (Eds.) *Constructivist views on the teaching and learning of mathematics* (pp. 65–78). Reston, VA: National Council of Teachers of Mathematics.

Davis, R. B., Maher, C. A., & Martino, A. M. (1992). Using videotapes to study the construction of mathematical knowledge by individual children working in groups. *Journal of Science Education and Technology, 1*(3), 177–189.

Devall, Y. (1983). Some cognitive and creative characteristics and their relationship to reading comprehension in gifted and nongifted fifth graders. *Journal for the Education of the Gifted, 5*(4), 259–273 (as cited in Ryser & Johnsen, 1998).

Dover, A., & Shore, B. M. (1991). Giftedness and flexibility on a mathematical set-breaking task. *Gifted Child Quarterly, 35,* 99–105.

Education Week on the Web. (2002, April 3). *Assessment.* Available: http://www.edweek.org/context/topics/issuespage.cfm?id=41.

Firestone, W. A., Schorr, R. Y., & Monfils, L. (2004). *The ambiguity of teaching to the test.* Mahwah, NJ: Erlbaum .

Goldin, G. A. (in press). Aspects of affect and mathematical modeling processes In R. Lesh, E. Hamilton & J. Kaput (Eds.) *Real-world models and modeling as a foundation for future mathematics education.* Mahwah, NJ: Erlbaum.

Goldin, G. A. (2002). Affect, meta-affect, and mathematical belief structures. In G. C. Leder, E. Pehkonen, & G. Törner (Eds.). *Beliefs: A hidden variable in mathematics education?* (pp. 59–72). Dordrecht, The Netherlands: Kluwer.

Goodkin, S. (2005, December 27). Leave no gifted child behind. *The Washington Post.* Retrieved June 28, 2006, from http://www.washingtonpost.com.

Greenes, C. (1981). Identifying the gifted student in mathematics. *Arithmetic Teacher. 28*(6). 14–17.

Hiebert, J., & Wearne, D. (1993). Instructional tasks, classroom discourse, and students' learning in second grade arithmetic. *American Educational Research Journal, 30*(2), 393–425.

House, P. A. (Ed.). (1987). *Providing opportunities for the mathematically gifted, K–12.* Reston, VA: National Council of Teachers of Mathematics (as cited in Ryser & Johnsen, 1998).

Kaput J. J. (1998). Transforming algebra from an engine of inequity to an engine of mathematical power by "algebrafying" the K–12 curriculum. In *The nature and role of algebra in the K–14 curriculum: Proceedings of national symposium* (pp. 25–26). Washington, DC: National Academy Press.

Kaufman, J. C., & Baer, J. (2004). Sure, I'm creative—But not in mathematics: Self-reported creativity in diverse domains. *Empirical Studies of the Arts. 22*(2), 143–155.

Keitel, C., & Kilpatrick. J. (1998) Rationality and Irrationality of international comparative studies. In G. Kaiser, E. Luna, & I. Huntley (Eds.) *International comparisons in mathematics edcuation* (pp. 242–257). London: Falmer Press.

Marr, D., & Sternberg, R. (1986). Analogical reasoning with novel concepts: Differential attention of intellectually gifted and nongifted children to relevant and irrelevant novel stimuli. *Cognitive Development, 1,* 5–72 (as cited in Ryser & Johnsen, 1998).

Miechenbaum, D. (1980). A cognitive-behavioral perspective on intelligence. *Intelligence, 4*(4), 271–283 (as cited in Ryser & Johnsen, 1998).

Miller, R. C. (1990). *Discovering mathematical talent.* Reston, VA: Council for Exceptional Children. (ERIC Document Reproduction Service No. ED.321 487 (as cited in Ryser & Johnsen, 1998).

National Council of Teachers of Mathematics. Curriculum and evaluation standards for school mathematics. Reston, VA.: The Council, 2000.

NCES: http://nces.ed.gov/nationsreportcard/

Niss, M. (1996). Goals of mathematics teaching. In A. Bishop, et al. (Eds.), *International handbook of mathematics education* (pp. 11–47). Dordrecht: Kluwer Academic Publishers.

O'Connor, M., & Hermelin, D. (1979). Intelligence differences and conceptual judgment. *Psychological Research, 41,* 91–100.

Reynolds, S. (2005). A study of fourth-grade students' explorations into comparing fractions. (Doctoral dissertation, Rutgers, The State University of New Jersey at New Brunswick 2005) Dissertation Abstracts International 66/04. p. 1305. AAT 3171003.

Ryser, G. R., & Johnsen, S. K. (1998). *Test of mathematical abilities for gifted students: Examiner's manual.* Austin, TX: Pro-ed.

Schmidt, W. H., McKnight, C. C., & Raizen, S. A. (1996). *A splintered vision: An investigation of U.S. science and mathematic education.* East Lansing, MI: U.S. National Research Center for the Third International Mathematics and Science Study.

Schorr, R.Y., & Bulgar, S. (2003). The impact of preparing for the test on classroom practice. In N. A. Pateman, B. J. Dougherty, & J. T. Zilliox (Eds.), *27th Conference of the International Group for the Psychology of Mathematics Education held jointly with the Twenty-fifth Conference of the North American Chapter of the International Group for the Psychology of Mathematics Education: Vol. 4. Navigating Between Theory and Practice* (pp. 135–142). Honolulu, HI: CRDG, College of Education, University of Hawai'i.

Schorr, R.Y., & Goldin, G. A. (in preparation). Affect and motivation in the SimCalc classroom. *Educational Studies in Mathematics.*

Scruggs, T., & Mastropieri, M. (1984). How gifted students learn: Implications from recent research. *Rooper Review, 6,* 183–185 (as cited in Ryser & Johnsen, 1998).

Scruggs, T. Matrropieri, M., Monson, J., & Jorgensen, C. (1985). Maximizing what gifted kids can learn: Recent finds of learning strategy research. *Gifted Child Quarterly, 29*(4), 181–185 (as cited in Ryser & Johnsen, 1998).

Shore, B. (1986). Cognition and giftedness: New research directions. *Gifted Child Quarterly, 30,* 24–27 (as cited in Ryser & Johnsen, 1998).

Smith, M. S., & Stein, M. (1998). Selecting and creating mathematical tasks: From research to practice. *Mathematics Teaching in the Middle School, 3*(5), 344–350.

Sriraman, B. (2003). Mathematical giftedness, problem solving, and the ability to formulate generalizations. *The Journal of Secondary Gifted Education, 14*(3), 151–165.

Sriraman, B. (2004). Discovering a mathematical principle: The case of Matt. *Mathematics in School, 33*(2), 25–31.

Sriraman, B. (2005). Are mathematical giftedness and mathematical creativity synonyms? A theoretical analysis of constructs. *Journal of Secondary Gifted Education, 17*(1), 20–36.

Stanley, J. C. (1976). The study of mathematically precocious youth. *Gifted Child Quarterly, 26,* 53–56. (as cited in Ryser, & Johnsen, 1998).

Steencken, E. P. (2001). Studying fourth graders' representations of fraction ideas. (Doctoral dissertation, Rutgers, The State University of New Jersey, New Brunswick, 2001) Dissertation Abstracts International 62/03. p. 953, AAT 3009381.

Stein, K. S., Smith, M. S., Henningsen, M. A., & Silver, E. A. (2000). *Implementing standards-base mathematics instruction: A casebook for professional development.* New York: Teachers College Press.

Sternberg, R. J. (1982). A componential approach to intellectual development. In R. J. Sternberg (Ed.) *Advances in the psychology of human intelligence* (Vol. 1, pp. 413–466). New York: Wiley (as cited in Ryser & Johnsen, 1998).

Sternberg, R. J., & Powerll, J. (1983). The development of intelligence. In J. H. Flavell & E. M. Markham (Eds.) *Handbook of child psychology: Cognitive development.* (Vol. 3, pp. 341–419). New York: Wiley (as cited in Ryser & Johnsen, 1998).

Towers, J. (1998). *Teacher's interventions and the growth of students' mathematical understanding.* Unpublished doctoral dissertation. The University of British Columbia, Canada.

U.S. Commission on National Security for the 21st Century. February 15, 2001. *Roadmap for national security: Imperative for change.* Washington, DC: GPO, Chapter II, pp. 30–46.

Von Rotz, L., & Burns, M. (2002, Fall). Caterpillars—A lesson with first graders. *Math Solutions online newsletter. (7).* Retrieved Fall 2002 from http://www.mathsolutions.com .

Warner, L., & Schorr, R. Y. (2004). From primitive knowing to formalizing: The role of student-to-student questioning in the development of mathematical understanding. *Twenty-sixth Conference of the North American Chapter of the International Group for the Psychology of Mathematics Education: Building Connections Between Communities.* Toronto, Ontario.

Wong, B. (1982). Strategic behaviors in selecting retrieval cues in figted, normal achieving and learning disabled children. *Journal of Learning Disabilities. 13,* 33–37 (as cited in Ryser & Johnsen, 1998).

Woodrum, D. (1975). A comparison of problem-solving performance for 4th, 5th and 6th grade children classified as normal, gifted, or learning disabled and by focusing level and conceptual tempo. *Dissertation Abstracts International 39*(11-A)6708–6709 (as cited in Ryser & Johnsen, 1998).

ACKNOWLEDGMENT

Reprint of Bulgar, S. (2007). Designing opportunities for all students to demonstrate mathematical prowess. In B. Sriraman (Guest Editor), *Mediterranean Journal for Research in Mathematics Education,* 6(1 & 2), 103–126. Reprinted with permission from The Cyprus Mathematical Society. ©2007 Sylvia Bulgar.

NOTE

1. This problem was created by Mr. Robert Krech, K–3 Math Research Specialist in West Windsor-Plainsboro Public School District in Central New Jersey. He was also the facilitator of the mentioned workshop.

CHAPTER 7

PROBLEMS TO DISCOVER AND TO BOOST MATHEMATICAL TALENT IN EARLY GRADES

A Challenging Situations Approach

Viktor Freiman
Université de Moncton, Canada

ABSTRACT

Several studies of mathematical giftedness conducted in the past two decades reveal the importance of creation of learning and teaching environment favourable to the identification and nurturing mathematically talented students. Based on psychological, methodological and didactical models created by Krutetskii (1976), Shchedrovtiskii (1968), Brousseau (1997) and Sierpinska (1994), we have developed our challenging situation approach. During 7 years of field study in the elementary K–6 classroom, we collected sufficient amount of data that demonstrate how these challenging situations help to discover and to boost mathematical talent in very young children keeping and increasing their interest towards more advanced mathematics curriculum. In

Creativity, Giftedness, and Talent Development in Mathematics, pages 155–184

155

this article, we are going to present our model and illustrate how it works in the mixed-ability classroom. We will also discuss different roles that teachers and students might play in this kind of environment and how each side could benefit from it.

INTRODUCTION

The biographers of famous mathematicians often refer to the evidence of a particular nature of their talent which can be detected already at a very young age. One can ask where this deep insight in mathematics comes from. How can teachers discover their talent and nurture it? And, as a result of this discovery, what kind of a classroom environment would be advantageous for these children? What can be done by teachers to help these children to realise their potential?

From their very early pre-school and school years, mathematically gifted children are active and curious in their learning, persistent and innovative in their efforts, flexible and fast in grasping complex and abstract mathematical concepts, and thus represent a unique human intellectual resource for our society, which we have no right to waste or to loose.

Numerous studies of mathematical giftedness conducted during past decades provide us with different lists of characteristics of gifted children and suggest various models of identification and fostering them in and beyond mathematics classroom.

Long time experimentation with schoolchildren and observations made by teachers allowed Krutetskii (1976) to construct a list of characteristics of mental activity have shown by mathematically gifted children in a comparatively early age:

- An ability to generalize mathematical material (an ability to discover the general in what is externally different or isolated)
- A flexibility of mental processes (an ability to switch rapidly from one operation to another, from one train of thought to another)
- A striving to find the easiest, clearest, and most economical ways to solve problems
- An ability chiefly to remember generalized relations, reasoning schemas, and methods of solving type-problems
- Curtailment of the reasoning processes, a shortening of its individual links
- Formation of elementary forms of a particular 'mathematical' perception of the environment—as if many facts and phenomena were refracted through prism of mathematical relationships.

Miller (1990) mentions some other characteristics that may give important clues in discovering high mathematical talent:

- Awareness and curiosity about numerical information
- Quickness in learning, understanding and applying mathematical ideas
- High ability to think and work abstractly
- Ability to see mathematical patterns and relationships
- Ability to think and work abstractly in flexible, creative way
- Ability to transfer learning to new untaught mathematical situations

Another model focusing on giftedness as "intersection" of various factors has been developed by Renzulli (1977). By means of this model, Ridge and Renzulli (1981) define giftedness as an interaction among three basic clusters of human traits: above average general abilities, high levels of task commitment, and high levels of creativity. Upon their definition, gifted and talented children are those possessing or capable of developing this composite set of traits and applying them to any potentially valuable area of human performance.

In a similar way, Mingus and Grassl (1999) focused their study on students who display a combination of willingness to work hard, natural mathematical ability and/or creativity.

The authors consider *natural mathematical ability*, which might be represented by several characteristics discovered by Krutetskii (see above) as well as non-mathematical ones as *willingness to work hard* (that means being focused, committed, energetic, persistent, confident, and able to withstand stress and distraction) or *high creativity* (i.e., capacity of divergent thinking and of combining the experience and skills from seemingly disparate domains to synthesise new products or ideas). The authors labelled students possessing a high degree of mathematical ability, creativity, and willingness as "truly gifted."

Reflecting on our classroom observations of 4–5-year old children using educational software with some mathematical tasks we became interested in studying deeply mathematically precocious children

We noticed that some of them always choose more challenging activities, go through all the levels up to the highest ones, understand each activity almost without any explanation from the teacher, demonstrate very systematic approach to the problem, have very sharp selective memory of important facts, details, methods, they are very creative in their work with "open-ended" problems (such as creating puzzles and patterns), and often share their discoveries with their peers being very proud of themselves.

For example, working with counting tasks such as finding a domino piece with number of dots corresponding to a show number from 6 to 9, some children count all the dots on almost *every* card using their fingers, others choose first one which contains *more* than 5 dots (for example, they may choose 8), then again, most of them count dots and if the result is not good, they jump randomly to another with similar number of dots. There is also a small group of children that try to spot a card with less than 8 dots on it. Finally, one child clicks immediately on card with 7 dots, saying "I know it's this one because 5 and 2 make 7."

Analyzing children's strategies, we could see their *different approach to numbers*. Some children see cards as pictures with objects to count and they use the same strategies as they were manipulative objects (like toys). Other children try to use a different, more complex approach—thinking globally (I see it's five here, I know that 7 is less than 8) and abstractly one (number as an abstract characteristic of a set of dots) along with using a number of shortcuts which helped them to increase efficiency of their mathematical work.

Our next example is a comparison task with two cards shown to the child: one with a certain number of dots arranged within an array 3×4 (12 dots as maximum) and another one with a number 1–12 written on it as a digit. The child has to decide *whether two cards present the same numbers or not*. For the most of 5 year old children, this is a relatively simple task but adding the time limit does make activity extremely challenging for children whose strategy of counting is limited by "finger pointing." The best winning strategy was found by children who used estimation (I know that I have much more dots here than number 3 on another side) and counting with eyes (without fingers). Some children were giving surprisingly deep comments like "I know this number of dots is 12 because I see 4 row of 3 dots which make 12" which demonstrate precocious insight into numbers and number relationships.

Some tasks give children an opportunity to create some patterns asking to construct a personage following certain pattern, or to create their own personage. This second option was seeing by many children more as an art activity, although our observation showed that some 4-year-old children create personages upon more complex pattern of mathematical nature (like color, background, part of cloths). One activity presented a grid 6×6 with a set of different puzzles to reproduce (pictures are given as a model) or to create their own puzzle and many young children (4–5 year old) did it just as drawing another picture.

Again, we could notice few children building spontaneously more mathematically abstract tessellations using complex, sometimes symmetrical configurations of shapes which would be more expected from older children already familiar with geometric transformations like reflection or translation. Another activity presented a factory for making cookies with chocolate chips on them. One mode of this activity asks child to put a number of chips on a cookie corresponding to a randomly given number (1 to 10). Another mode prompts to create a cookie with an arbitrary chosen number of chips. Giving free choice to children it allowed us to observe some of them making cookies with consecutively chosen numbers from one to ten repeated in two rows. And even more, they were so fascinated with their result so they started to repeat the same pattern more and more without any visible fatigue, although it was a routine repetition of the same procedure. It seems that here we have an example of a mathematical creativity of a particular kind: seeing beauty of mathematical structure in the same repeating pattern.

Equalizing task is a complex task for very young children. For example, an activity of feeding rabbits with carrots shows some rabbits "waiting for a food", on another—an empty field in which a child has to put carrots keeping in mind that each rabbit would get one carrot. In fact, the child has to control two conditions at the same time to ensure that number of rabbits is equal to the number of carrots. Our observation shows that some children decide to arrange carrots in a certain geometric pattern (row, stair, or array) helping themselves to keep control of conditions showing thus more complex way of thinking.

Finally, working on ordering tasks like one of arranging 7 dolls "matreshka" in increasing or decreasing order by size, some children proceed rather by trial and error, others do it more systematically (looking at neighbors and switching if necessary). Few of them do it in a very systematical way: starting with putting a smallest/biggest one first, then going to the next smallest/biggest and so on. This strategy allows them to simplify the process of problem solving, and at the same time, shows their ability to apply more complex thinking.

Reflecting on these examples, one can ask: Why do these children demonstrate such unusual behavior at an early age? Is it simply related to the attractiveness of computer games on the screen, or does it reflect a much more complex structure of their mind? Our further study of these children's strategies while solving "purely" mathematical tasks led us to believe that, indeed, the latter might be the case and that it is worth while searching for a specific structure of the mathematically able mind.

Our further questions were: How to identify "pure" mathematical components of the children's learning activity? What kind of cognitive structure enables a child to act like a mathematician? And, from the point of view of practicing teacher, we asked: How to organize children's mathematical activities so that they were motivated to act this way?

In the following section, we will analyze several theories that form our theoretical framework enabling us to analyze problems that help to boost mathematical talent in young children.

THEORETICAL BACKGROUND

Kulm (1990) remarks that since so much of school mathematics in the past has been focused on practised skills, the completion of a large number of exercises in a fixed time period has been accepted not only as a measure of mastery, but as an indication of giftedness and potential for doing advanced work. On the other hand, higher order thinking in mathematics is by very nature complex and multifaceted, requiring reflection, planning, and consideration of alternative strategies. Only the broadest limits on time for completion make sense on a test purposing to assess this type of thinking.

Burjan's (1991) recommendation to use

- Open-ended investigations and open-response problems rather than multiple-choice
- Problems allowing several different approaches
- Non-standard tasks rather than standard ones
- Tasks focusing on high-order-abilities rather than lower-level-skills
- Complex tasks requiring the use of several "pieces of mathematical knowledge" from different topics) rather than specific ones (based on one particular fact or technique)
- Knowledge-independent tasks rather than knowledge-based one goes in the same direction.

Unfortunately, as it was mentioned by Greenes (1981), the bulk of our mathematics program is devoted to the development of computational skills and we tend to assess students' ability or capability based on successful performance of these computational algorithms (so called "good exercise doers") and have little opportunity to observe student's high order reasoning skills.

Sometimes, even a very banal math problem might deliver a clear message about distinguishing the gifted student from the good student. Greenes analyses a very simple word problem (given to 5th grade children):

Mrs. Johnson travelled 360 km in 6 hours. How many kilometres did she travel each hour?

One bright student surprised the teacher by having difficulty to solve this easy problem. Finally, the teacher realised that the student has discovered that nothing was said about the same number of kilometres travelled each day. This example demonstrates the child's ability to detect ambiguities in the problem, which indicate him/her as mathematically gifted student.

That is why, in a later work Greenes (1997) insists on the importance of presenting situations in which students can demonstrate their talents: "One vehicle for both challenging students and encouraging them to reveal their talents is to use of rich problems and projects." Greenes mentions that such problems accomplish the following:

- Integrate the disciplines (application of concepts, skills, and strategies from the various sub-discipline of mathematics or from other content areas (including non-academic ones)
- Are open to interpretation or solution (open-beginning and open-ended problems)
- Require the formation of generalisations (recognition of common structures as basic to analogue reasoning)
- Demand the use of multiple reasoning methods (inductive, deductive, spatial, proportional, probabilistic, and analogue)
- Stimulate the formulation of extension questions
- Offer opportunities for firsthand inquiry (explore real-word problems, perform experiments and conduct investigations and surveys)
- Have social impact (well-being or safety of members of the community)
- Necessitate interaction with others

Many authors point at teacher's particular role in the process of identification of mathematically able children. Kennard (1998) affirms that the nature of the teacher's role is critical in terms of facilitating pupil's exploration of challenging material. Hence, the identification of very able pupils becomes inextricably linked with both the provision of challenging material and forms of teacher-pupil interaction capable of revealing key mathematical abilities. The author votes for interactive and continuous model for providing identification through challenge which integrates the following strands:

- The interpretative framework
- The selection of appropriately challenging mathematical material
- The forms of interaction between teachers and pupils which provide opportunities for mathematical characteristics to be recognised and promoted
- The continuous provision of opportunities for mathematically able children to respond to challenging material

In Kennard's case study based on this model and Krutetskii's categories the identification was conducted by the so-called teacher-researcher in the classroom environment where the pupils are being taught as well as observed. The questioning approach was used in order to reveal aspects of pupils' mathematical approaches and understanding.

Ridge and Renzulli (1981) suggest three types of activities which are important for nurturing mathematical talents:

- General exploratory activities to stimulate interest in specific subject areas: experiences that would demonstrate various procedures in the professional or scientific world (through children's museums and science centres) in which students would get an opportunity to choose, explore, and experiment without the treat of having to prepare report or provide any sort of formal recapitulation.
- Group training activities to develop processes related to the areas of interest developed through general activities. The aim of these activities is to enable students to deal more effectively with content through the power of mind. Typical for these thinking and feeling processes are critical thinking, problem solving, reflective thinking, inquiry training, divergent thinking, sensitivity training, awareness development, and creative or productive thinking. Problem solving applies to:
    1. The application of mathematics to the solution of problems in other fields
    2. The solution of puzzles or logically oriented problems
    3. The solution of problems requiring specific mathematical content and processes.
- Individual and small-group investigation of real problems. As giftedness becomes manifest as result of student's willingness to go engage in more complex, self-initiated investigative activities, the essence of this type of activities is that students become problem finders as well as problem solvers and that they investigate a real problem using methods of inquiry appropriate to the nature of the problem (p. 231).

For his study, Krutetskii (1976) developed several sets of challenging mathematical problems and conducted interviews with each of chosen students offering an original way to study mathematical abilities within appropriate mathematical activity, which, taken in school instruction, consists of solving various kind of problems in the broad sense of word, including problems on proof, calculation, transformation, and construction. He analyses seven principles of a choice of mathematical problems suitable to discover mathematically able student:

1. The problems should represent about equally the different parts of school: mathematics-arithmetic, algebra, and geometry
2. Experimental problems should be of various degrees of difficulty

3. They ought to fulfil their direct purpose: solving them should help to clarify the structure of abilities
4. The problems should be oriented not so much toward a quantitative expression of the phenomenon being studied as toward revealing its qualitative features (process versus result)
5. We should try to choose problem the solving of which is primarily based on abilities, not on the knowledge, habits, or skills
6. The problems have to allow to determine how rapidly a pupil progressed in solving problems of a certain type, how well he achieved skill in solving these problems, and what were his maximum possibilities in this regard (instruction versus diagnostic)
7. The problems are supposed to allow some quintile analysis as well as qualitative one.

Analyzing different children's approaches to the problems, Krutetskii (1976) provides us with several key elements of mathematical ability showing how these challenging problems help us to recognize different children's approaches to mathematics. In a regular classroom, we often teach students direct methods of solving mathematical problems. Then, in order to test their knowledge, we give them the same kind of problem and expect them to (re-)produce the same solution.

This might lead to several paradoxes, such as Brousseau's (1997) paradox of devolution of situations when the teacher "is induced to tell the student how to solve the given problem or what answer to give, the student, having had neither to make a choice nor to try out any methods nor to modify her own knowledge or beliefs, will not give the expected evidence of the desired acquisition." Brousseau thus claims that everything the teacher undertakes in order to make the student produce the behaviors that she expects tends to deprive this student of the *necessary conditions for understanding and learning of the target notion.*

However, several studies point at the fact that, in order to access a higher level of knowledge or understanding, a person has to be able to proceed at once with an integration and reorganization (of previous knowledge). Sierpinska (1994) sees the need of "reorganizations" as one of the most serious problems in education. But we can not just tell the students "how to reorganize" their previous understanding, we can not tell them what to change and how to make shifts in focus or to generalize because we would have to do this in terms of knowledge they have not acquired yet.

Looking for new methodological approaches to teaching and learning, Shchedrovitskii (1968) gives striking examples of other paradoxes when we as educators want our children to master some kind of action by teaching it directly giving children tasks which are identical with this action. But classroom practice shows that the children not only do not learn actions that go beyond the tasks, they do not even learn the actions that we teach them within the tasks.

In our challenging situation model, we propose an active everyday use of open-ended mathematical activities that would engage children into a meaningful process of exploring, questioning, investigating, communicating and reflecting on mathematical structures and relationships. This model represents rather larger vision of mathematical giftedness that correlates with Sheffield's notion of a mathematical promise (Sheffield, 1999) and thus aims to give pleasure to more children to think and to act in a mathematically meaningful way.

OUR STUDY: GENERAL CONTEXT

Our experiment reflects 7 years of classroom activities and observations with Grades K–6 children while teaching challenging mathematics courses. It has been conducted at one Montreal located private bilingual elementary school with French and English both taught as a first language. Along with a strong linguistic program (with a third language, Spanish or Italian), the school insists on offering enriched programs in all subjects including mathematics to all its students independently of their abilities and academic performance.

The school thus promotes education as a fundamental value by instilling the will to learn while developing the following intellectual aptitudes:

- Being able to analyse and synthesize
- Critical thinking
- Art of learning

The mathematics curriculum is composed of a solid basic course whose level is almost a year ahead in comparison to the program of the Quebec's Ministry of Education (Programme de formation de l'école québécoise, 2001) and an enrichment (deeper exploration of difficult concepts and topics: logic, fractions, geometry, numbers as well as a strong emphasis on problem solving strategies). The active and intensive use of "Challenging mathematics" text-books (Lyons & Lyons) along with carefully chosen additional materials helps us create a learning environment in which the students participate in decisions about their learning in order to grow and progress at their own pace. Each child competes with himself (herself) and is encouraged to surpass himself (herself).

Since the school doesn't do any selection of students for the enriched mathematics courses, all its students (total of 238) participated in the experiment. With some of them, this author started to work at their age of 3–5, as a computer teacher. There were many students that we could observe during a long period of time (for example some of Grade 6 children in 2002–2003 were our students since Grade 1, some of them since the age of 3–5). During this period, some children had to leave the school, some of them joined the class later (in the same Grade 6, there were 2 students who started in our school in Grade 6). In terms of abilities, we can characterise our classroom as a mixed ability classroom with a significant variation in the level of achievement.

The enriched course aimed to foster children's logical reasoning and problem solving skills in all children. It is based on *challenging situations* presented in the 'Challenging Mathematics' textbook collection (Défi mathématique (Lyons, Lyons)) along with other different computer and printed resources (LOGO, Cabri, Game of Life, Internet, and so on) as well as situations created by the author. It included several topics earlier than in the regular curriculum; some topics were presented in more depth than in the regular curriculum; various topics which are not included in the regular curriculum. Such curriculum thus requires a mobilising of all the inner resources of the child: her motivation, hard mental work, curiosity, perseverance, thinking ability. Since all our students are exposed to this enriched curriculum, the differences between them become more evident.

In our further more detailed analysis, we will make explicit the role of the challenging situation itself showing that without the context of challenging situation, such opportunity for students and teachers would be lost.

CHALLENGING TASKS AS POWERFUL TEACHING AND LEARNING TOOLS HELPING TO DISCOVER AND BOOST MATHEMATICAL TALENT

The story of Gauss solving a routine problem of calculating the sum of the first hundred natural numbers is one of the well known examples of this kind. While all other children were desperately trying to add terms one by one, Gauss impressed the teacher by finding a quick and easy way to do it regrouping the terms in a special way (see, for example, Dunham, 1990).

But one can ask: what were the characteristics of the classroom situation, which allowed the gifted student to demonstrate his talent in mathematics?

The same story says that the teacher had chosen the task for its accessibility to all students (the task is routine) and the probably very long time that it would take the students to solve; he hoped to thus keep them all quiet and busy for a good while. What he didn't expect that one of the students would turn the routine task into a challenging one of finding a quick way of solving an otherwise tedious and long computational exercise. The situation was not planned to reveal a mathematical talent, yet it did so "spontaneously." The situation became a challenging one by chance.

In many similar cases, mathematical talent would not be identified. We could say that using routine drill tasks involving numerous standard algorithms is not, in general, offering a good opportunity to identify and nurture mathematical talents.

Sheffield (1999) calls such routine tasks "one dimensional." As an example, she cites a class of three-four graders reviewing addition of two-digit numbers with regrouping. Children are asked to complete a page of exercises such as: 57 + 45, 48 + 68, 59 + 37. As it usually happens with brighter and faster students, they finish all exercises before their classmates. So the teacher would "challenge" them with 3- or 4-digit additions. Although the calculations become longer and time consuming, the tasks themselves are not more complex or more mathematically interesting.

As a better didactical solution for these children, Sheffield suggests the use of meaningful tasks like one of *finding three consecutive integers with a sum of 162*:

> Students would continue to get the practice of adding two-digit numbers with regrouping, but they also would have the opportunity to make interesting discoveries. Students who are challenged to find the answer in as many ways as possible, to pose related questions, to investigate interesting patterns, to make and evaluate hypotheses about their observations, and to communicate their findings to their peers, teachers, and others will get plenty of practice adding two digit numbers, but they will also have the chance to do some real mathematics. (Sheffield, 1999: 47)

By giving children a challenging task we would expect them to make efforts in understanding a problem, to search for an efficient strategy of solving it, to find appropriate solutions and to make necessary generalizations.

Following examples illustrate three very different approaches to the same problem of *finding a number of handshakes that we obtain when* n *people shake hands of each other* used by mathematically talented children.

Marc-Etienne (10) organized an experience with his classmates, considering systematically the cases $n = 2$, $n = 3$, etc. Then he made then necessary generalizations. Here is a transcript of his report in which we could observe several steps:

Step 1: two circles connected with an arrow representing two people – one handshake. He wrote beside the picture "= 1."

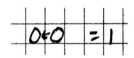

Step 2: three circles forming a triangle connected with three arrows representing three people—three handshakes. He wrote beside the picture "= 3"

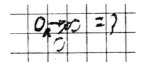

Step 3: four circles forming a square connected with six arrows representing four people – six handshakes. He wrote aside: "3 + 2 + 1 = 6" commenting:

"1 after another, they leave"

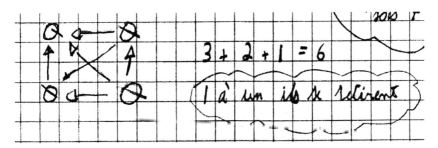

Step 4: Five circles form a "domino–5-dots disposition" connected with only six arrows (some arrows are missing). However, he wrote: "4 + 3 + 2 + 1 = 10" continuing the same pattern.

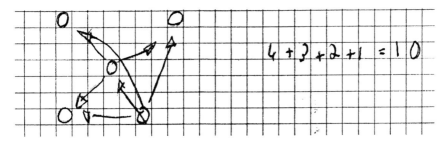

Step 5: Six circles disposed in two rows (by three), no arrows. He wrote: "5 + 4 + 3 + 2 + 1 = 15"

Step 6: Seven circles disposed in two rows (three + four), no arrows. He wrote: "6 + 5 + 4 + 3 + 2 + 1 = 21"

Step 7: Eight circles disposed in two rows (by four), no arrows. He wrote: "7 + 6 + 5 + 4 + 3 + 2 + 1 = 28"

He concluded his generalization with following sentence that he called:

Formule: *On calcule toujours de la manière que le prochain chiffre soit –1 2 + 1 et que chaque chiffre qui précède soit +1* (We calculate always in the same way in order that each next number in our sum would be 1 less. 2 + 1 and each previous number would be +1).

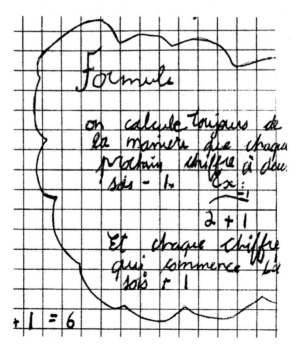

Charlotte (10) used a particular case of 5 people making diagrams doing systematic search for all possible combinations.

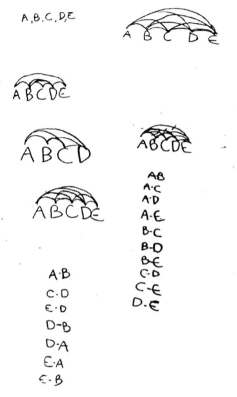

Christopher (10) was very short in his presentation writing just one single sentence:

$$1 + 2 + 3 + 4 + 5 + 7 + 8 = 36$$

He provided it with the oral explanation that if we have a group of people, each person has to shake hands to all people who came before, so with 2 people we would have 1 handshake, with 3 people –2 more handshakes (1 + 2), and so on.

We can see that investigation of initial situation of 'handshakes' allowed children to explore the problem, to look for patterns and to make important mathematical generalizations. Moreover, with a simple boosting question what would be the number of handshakes with 101 people, the class arrived to the same problem that Gauss had to deal with but posed in different, challenging way. Thus, a further investigation might be provoked here in a more natural way.

And, from the point of view of practicing teacher, we can better understand *how to organize children's mathematical activities so that they were motivated to act this way.*

FROM CHALLENGING TASKS TO CHALLENGING CURRICULUM: EXAMPLE OF KINDERGARTEN ENRICHMENT COURSE

There are two basic approaches to design a mathematics curriculum for 5–6 year old children; one can be labelled as traditional and the other as innovative. The former is based on counting, ordering, classifying, introduces basic numbers, operations (addition and subtraction), relations (more, less, bigger, smaller, greater) and shapes. The latter puts more emphasis on learning while allowing children to play using manipulatives, colouring, arts and crafts, games with numbers and shapes. During the past decade, many creative teachers have been trying to use the best ideas from each of the two approaches also adding reasoning activities to the mathematics curriculum.

In our school, we used a traditional approach based on Quebec's *Passeport Mathématique* Grade 1 textbook along with a new French collection *Spirale (Maths CP2)* which represents the second, modern approach. However, even this combination doesn't provide our children with the material necessary for their mathematical development. There is still a gap between their level of ability and the requirements of the challenging curriculum that we use starting from Grade 1 (collection "Défi mathématique") and which is based on discovery, reasoning and understanding.

In order to fill the gap, we developed an enriched course offered to all kindergarten students (we have 30–35 children every year). The course was given on a weekly basis (1 hour a week). We base our teaching on the challenging situations approach, developing activities that stimulate mathematical questioning and investigations along with reflective thinking.

Each class starts with such questions as *What did we do last time? What problem did we have to solve? What was our way to deal with the problem? What strategies did we use?*, etc. This questioning aims to provoke reflection on the problems that children solved as well as on methods that they used. Without this reflection, rupture situation (in Shchedrovitskii's sense) would never arise, because a rupture is a break with previous knowledge, which needs to be brought to mind.

At the same time, we would ask questions that would indicate children's understanding of underlying mathematical concepts or methods that we aim to introduce (using appropriate vocabulary and/or symbolism).

During this initial discussion we usually try to bring in a new aspect which provides children with an opportunity to ask new questions, to look at the problem in a different way. Sometimes, we might ask them, simply, *what do they think we should do today?*

Thus we can pass to the new situation/new problem/new aspect of the old problem. We may do it by means of provoking questions, of interesting stories or introductory games. Following Shchedrovitskii and Brousseau, we try to avoid the teaching paradox by not providing children with direct description of the tasks or methods of solutions. We try also to keep their attention and motivate them.

After this introductory stage, children begin investigating a problem using different manipulatives: cubes, geometrical blocs, counters, etc. They work alone or in groups. During the phase of investigation, the role of the teacher becomes more modest: we give children certain autonomy to get familiar with the problem, to choose a necessary material, organise their work environment, and choose an appropriate strategy.

However, some work has to be done by the teacher to guide children through their actions. We have to make sure that the child understands the problem, the conditions that are given (rules of the game), the goal of the activity. As the child moves ahead, we shall verify his control of the situation: what she is doing now and what is the purpose of the action? (activating reflective action). We have to keep in mind that the exploration is used not only as a way to make the child do some actions but also and foremost as an introduction to mathematical concepts or methods.

Therefore, the teacher needs to be prepared to introduce the necessary mathematical vocabulary along with its mathematical meaning as well as mathematical methods of reasoning about the concepts and about the reasoning. In our experiment, we try to choose those mathematical aspects that are considered as difficult and are not normally included in the Kindergarten curriculum.

For example, when we want to introduce an activity with patterns, we would organise a game. We would start to make a line "boy, girl, boy, girl,..." children find it easy and are happy to discover a pattern. Then we would start a new "pattern": "boy, girl, boy, girl, boy, boy." Many children would protest, saying that the pattern is wrong. But perhaps, some of them would try to look for different pattern, like "glasses, no glasses, glasses, no glasses,..."

As the game goes on, children get used to looking for familiar patterns. This is the time to challenge them more. For example, we may ask them, how many children would be in the line with the pattern "boy, girl, boy, girl,..." Since there were only 8 boys in the classroom, one child could make a hypothesis that it gives 8+8 children in the line. After such a line had been completed, teacher's silence could be broken by a child's voice— "we can add one more child to the line—a girl in the beginning."

The course is built of various challenging situations that we create in order to give children an opportunity to take a different look at mathematical activities that they usually do, to question their knowledge about mathematics trying to discover hidden links between different objects, to discover structures and relationships between data, learn to reason mathematically based on logical inference and at the same leave some space to children's mathematical creativity. We use different didactical variables in order to create obstacles making children re-organise their knowledge and create new means in order to overcome the obstacle. We were also asking our children to report on their investigations inviting them to communicate their discoveries by developing appropriate tools: diagrams, schemas, symbols, signs.

GUIDELINES FOR THE DESIGN OF CHALLENGING SITUATIONS

We consider three kinds of challenging situations:

- Open-ended problems and investigations
- Routine work turned into a challenge by the teacher
- Routine work turned into a challenge by a student

Let us consider these options in details:

Open-Ended Problems and Investigations

As we look at the video protocol of interviews with 4–6 year old children conducted by Bednarz and Poirier (1987) within their study of number acquisition by young children, we see how the evidence of differences in organisation of mathematical work by very young children becomes explicit with the open character of given tasks.

The video presents children's work on different tasks related to the concept of number: counting, formation of collections, order, conservation, comparison. Each task that in a regular classroom might be seen as ordinary, was given by authors in a very original challenging, dynamic, and open-ended way.

The child was constantly invited to think about the process of her work (how did you do it?), to develop an efficient strategy, to re-organise, if necessary, her process, to co-ordinate her actions. Thus, the routine tasks became open-ended and a child was given an opportunity to become an organiser of her mathematical work.

In our experiment, we also tried to make problems more open than they were usually presented to the students.

For example, we can take a problem from one mathematical competition:

$$\longrightarrow \quad \begin{array}{ccc} 1 & 2 & 3 \\ 4 & 5 & 6 \\ 7 & 8 & 9 \end{array} \longrightarrow$$

In this table, we enter by 1 and exit by 9.

One can only move horizontally or vertically, and it is impossible to step twice on one box. For example, moving through boxes 1–2–5–8–9, one gets a sum of 25. But not all the trajectories lead us to the number 25. Give all others 9 numbers.

This problem was given to participants of the regional final of the Championnat International des Jeux Mathématiques et Logiques in 2000 for Grade 4–5 children (10–11 year old) http://www.cijm.org/cijm.html.

We found that this problem would become more challenging for children if posed in a different way (open-ended):

Someone is going to visit a museum, which has 9 exhibition halls, arranged in a square 3×3. The number of paintings in each hall is written in the box. What are all the possible numbers of paintings that could be seen by this visitor who does not like to be in one hall twice ?

Not only do we hide the number of different ways, which makes this problem open, we give it to our Grade 1 students (6–7 year old). Every student had a task at his/her level (They will all be able to find at least a couple of solutions).

The following example illustrates the work of Chantal (6):

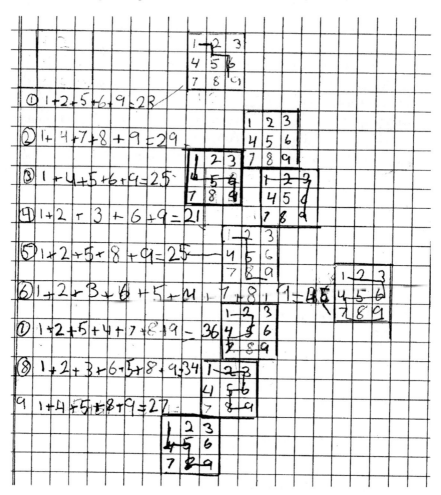

This example demonstrates how this open-ended situation helped the student to develop different abilities to organize systematic search and to keep tracks of her work

Routine Work Turned Into a Challenge by the Teacher

The recent Quebec's school curriculum puts emphasis on the importance of mastering basic number facts (like multiplication tables). This routine task can be made more challenging in many different ways. For example, one day we wrote on the board the 9-table operations:

$1 \times 9 =$
$2 \times 9 =$
$3 \times 9 = $, and so on.

Grade 3 children said immediately that it is a very easy table, because there is a well known regularity (writing down first digits of the product in order from 0 to 9 and the second ones down from 9 to 0, we obtain all the multiples of 9: 09, 18, 27, and so on). Among the answers one could find that $6 \times 9 = 54$.

So, the teacher comes to the board and writes $6 \times 9 = 56$ telling the story that when he was young, he had to memorise all answers, not just "tricks," and he is sure that $6 \times 9 = 56$. The students are confused, but many of them started to think how to prove that their result (54 was the correct one).

Many of them went to the board to share their ideas as well as other ways to obtain a 9-table. As a result of the lesson, the 9-table has appeared a couple of times on the board, children said it many times aloud, so they could memorise it and at the same time do it in a meaningful way questioning and proving their methods and ways of reasoning.

Routine Work Turned Into a Challenge by a Student

When Grade 4 children are asked to represent 1/8 of a rectangle, they find it an easy and routine task. That's why we were surprised by Christopher's way to divide a rectangle in 64 boxes (8 rows × 8 columns) and to colour 8 boxes randomly. He found that the task was not challenging enough and he wanted to make it more complicated.

Transformation of Challenge Within One Situation

All three ways of creating of challenging situations are not isolated from one another. They can also be transformed one into another.

For example, a kindergarten class (5–6 year old) is working on an *open-ended* problem:

> Amelie needs to build new houses for her farm animals. When one looks at the house from the sky, she sees that all of them have a roof in shape of a "digit." She has to build now a new house for her cows. What "digit" do you suggest to use for the roof of this new house?

Children used blocs in form of different solids. The activity aimed to make them to explore different solids, to make different constructions with them.

There are, basically, two ways of making constructions: three-dimensional or two-dimensional establishing thus different spatial relationships. For example, we may teach children to verify which shapes fit together recovering certain surface. The activity that we gave to our students didn't aim to teach any particular way of making constructions: many textbooks contain a lot of exercises asking children to reproduce one construction or another. Our situation was designed in order to help children get a certain "spatial feeling" trying different way to layout blocs. The main challenge in it was to organise a mathematically meaningful investigation within an "ill-defined" problem.

Some of them chose to imitate shapes of digits in the way we write them, others looked for different ways to create more economic constructions taking care of geometric properties (like seeing if the blocs fit one to another). Finally, there was a group of children who moved from the initially given situation of building a new house and started to construct many digits writing numbers (up to the 1000).

Soon, we could see that originally challenging and creative, the task became routine for many children. So, we decided to put some restrictions (new variables) that were sought as means to engage children in the investigation of a different problem in which we would be willing to construct house that has a "5"- shape and do this with a minimum of blocs. Thus, with the intervention of the teacher, a routine problem became a challenging one once again.

This method of a sudden change of didactic variable (Brousseau, 1997) is important in our study of relationship between child's organisation of the problem-solving and mathematical giftedness because it provokes a reflection (what is new?) and re-organisation of the whole process of thinking and acting (what do I need to modify?) and thus gives students a chance to show their full potential.

In our experimental work with young children, we obtained a constant confirmation of the fruitfulness of such an approach, especially if one wants to identify and nurture gifted children. David (5 years old) was working on the minimum task (See Figure 7.1); he looked happy with his solution (4 blocks) but still in what looked like a "state of alert." At this moment, we began to discuss children's solutions. One group of children has presented a three-blocs solution. Suddenly, David started to change something in his configuration; he lost completely "fiveness" of his shape while focusing on minimization task. But what is the most intriguing, is the rapid reactions of this child to the changing conditions (someone has found a better solution). This constant state of alert is an important characteristics of giftedness which could be better activated in challenging situation that in the ordinary one.

Figure 7.1

This state of "alert" leads them to constantly verify all the conditions going back and forth through the situation. Here is one more observation. Grade 4 students worked on their test. Answering a question of "Is it true that if the sum $a + c = 8$ then a and c are two different numbers?" Christopher hesitated a lot, saying however, that the numbers have to be different. As his work on the test went on, he had to solve a system of two equations with two variables: $ab = 16$, $a + b = 8$. He found easily $a = 4$, $b = 4$ as a solution then went back to his previous task and corrected his answer.

We could also observe another interesting phenomenon: challenging situation created by the teacher may initiate its further explorations by gifted students.

For example, doing the same activity with Grade 1 children (6–7 year old), we could state that it was seen as a routine problem by many of them and some of students lost completely their interest in it. Yet, we could still observe one girl looking for many different ways of building "5" using 4 blocs.

Not only she kept herself working on this problem, she came out with a new one: she started to look for possibilities to built a digit "4" with a minimum of blocks. Here, the problem was turned into a challenging one by the student.

The Role of the Teacher

In a challenging environment, the role of the teacher becomes crucial in all the stages:

- Choice of a problem
- Way of presenting it to the students
- Organisation of student's work
- Interpretation of results
- Follow-up

One of the very important conditions of success of the challenging situations approach is the teacher's attitude. How should we, as teachers, control the student's work? Related to the learning paradox (described in the previous chapter), it is far from being obvious how to find a solution to this problem. On the one hand, every word and every gesture said by teacher can affect the whole challenge of situation in either a positive or a negative sense. On the other hand, the teacher has to have a full didactical control of the situation (otherwise a mathematical learning activity might become a sort of 'arts and crafts in mathematical wrapping').

Our experiment didn't provide us with clear recipes but rather with examples that can be open to further questioning and investigations. These examples allowed us to formulate teacher's approaches favourable for the challenging situation:

- Give a child an opportunity to think: being a flexible teacher
- Support of children's willingness to learn more about math
- Challenge students in informal situations: sense of humour
- Support children in their desire to go beyond pre-planed situations
- Giving hints without telling solutions
- Management of particular cases of mathematical giftedness
- Use of little tricks as follows :

- While distributing manipulative material (blocs, cubes, etc.), we would give children time to touch it, to play with it, to get a feeling of it; sometimes it gives us important clues of children's organisations (how they put material, arrange it, order, classify, build different forms, etc.)
- When children finish their manipulation, we ask them to write a report. Sometimes it makes sense to give them time to break up their constructions. This opens the door to a variety of presentations (will the child reproduce his construction, add new details, draw a completely different pictures)
- When children are asked to communicate their results, it is important to motivate them to give detailed explanations. We often ask them to be mini-teachers—to explain to somebody who doesn't understand the problem
- Children often ask us to teach them complicated things. Sometimes, a pedagogical effect can be bigger if the teacher makes them wait. Then, starting to teach it, children might become more motivated: *finally, we got it!*

The Role of the Student

The role of the students in a challenging situation differs significantly from those in the regular learning activity. They have to adapt to a new, open environment. They have no precise algorithm of actions, no clear instruction what to do. Therefore, they have an opportunity to:

- Demonstrate different approaches to the problem
- Act differently in different situations
- Overcome obstacles, construct various means, discover new relationships
- Work on mathematical problems based on structures and systems using properties and definitions, conjectures and proofs
- Use of logical inference with fluency, control, rigour
- Combine logic and creativity in problem solving
- Invent new symbols and signs, use schemas and abstract drawings
- Use reflective thinking
- Ask mathematical questions, create new problems, investigate, use mathematics in non-mathematical situations, look around with "mathematical eyes"

CONCLUSIONS AND RECOMMENDATIONS

There are a number of educational studies of mathematical giftedness. Various models of giftedness based on different characteristics of mathematically gifted students have been developed and implemented. Different programs of support provide gifted students with advanced curriculum and guidance of highly qualified professionals. Several mathematical contests, Olympiads, and competitions help in searching for mathematically gifted children and taking care of their development.

Yet, the problems of identification and nurturing of mathematical talent are far from being solved. Many children become bored, at a very early age, with the simplified curriculum, lose their interest in mathematics and waste their intellectual potential. Despite the ingenious testing system, some children never get admitted to special programs for gifted students. The regular school system is not equipped to help these children.

Our study aimed to contribute to filling this gap and providing elementary school (Grades K–6) teachers with methods of identification and fostering mathematically gifted children in the mixed ability classroom.

We have called our approach, the "challenging situations approach." The approach is theoretically grounded in Krutetskii's (1976) notion of mathematical ability, Shchedrovitskii's (1968) developmental model of reflective action, Bachelard's (1938) notion of epistemological obstacle, Sierpinska's (1994) distinction between theoretical and practical thinking in mathematics, and Brousseau's (1997) theory of didactic situations.

Following Krutetskii (1976), we have defined mathematical ability as a "mathematical cast of mind," which represents a unique combination of psychological traits that enable young children to think in structures, to formalise, to generalise, to grasp relations between different concepts, structures, data and models and thus solve different mathematical problems more successfully than children of average or low ability.

At a very early age, these children demonstrate high thinking potential in reasoning about mathematical concepts and systems of concepts along with the capacity to reason about their reasoning. From the outset, they are better prepared than other children for theoretical thinking, which is the foundation of pure mathematical thinking.

The critical point of our study was an understanding that a discovery and nurturing of theoretical thinking is not possible if children are working with routine arithmetical tasks, merely applying algorithms that had been provided by the teacher, telling her students what to do and how to do it.

The paradoxes of such classroom situations have been described by Brousseau (1997) in his *Theory of Didactical Situations.* Following Brousseau's theory, we bring a notion of challenging situation into our model of mathematical giftedness postulating that a gifted child will show her talent in mathematics only in specific situations when a real question has been asked and a real problem has been posed.

"Challenging situations" use open-ended problems and mathematical investigations. A challenging situation initiates the student's action of structuring a problem, and of searching for links between data and with her previous experience. Since a real challenge is possible only when the situation is new for the learner, the challenging situation must contain a rupture with what the student has previously learned, provoking the student to reflect on the insufficiency of the past knowledge and construct new means, new mechanisms of action adapted to the new conditions, activating her full intellectual potential.

Challenging situation in its very nature gives many growing up opportunities for mathematical talent by:

- Providing the student with an opportunity to face an obstacle of a pure mathematical nature, the so called epistemological obstacle. In order to overcome it, the student will have to re-organise her mathematical knowledge, create new links, new structures following laws of logical inference. We claim that situations satisfying these conditions allow the teacher to identify and nurture mathematical giftedness among her students.
- Presenting a problem, which goes above or beyond the average level of difficulty. The child is encouraged to surpass what is normally expected of children of her age, thus demonstrating her precocity, which is a sign of mathematical giftedness.
- Helping to create a friendly environment in which a child compete with herself sharing her discoveries with other children and learning from others. Thus it gives mathematically gifted children who are not high achievers to participate actively in class and to succeed.

Challenging situation cannot be created as an isolated learning task. It full developmental potential can be realised only within a system of teaching based on a challenging curriculum as a whole. This would allow creating a learning environment in which every child would be able to demonstrate her highest level of ability.

This is why, using a challenging situation model we are not only able to get gifted children involved in genuine mathematical activity but also help all children to increase their intellectual potential.

Finally, challenging situation has another opening for gifted children: they can always go further, go beyond situations, ask new questions, initiate their own investigations, be more creative in their mathematical work. This spontaneous mathematical reaction feeds back into the learning environment in a positive way and further enhances its potential for all children.

We consider this feature of the approach as crucial from the point of view of mathematics education for all children.

Our study prompts different teaching approaches in mathematics. The teacher is no more re-translator of knowledge or instructor of methods of problem solving. In a challenging situation her role becomes more as moderators of discussions, listeners of student's ideas, student's guide through the discovery.

In helping students go through various obstacles, we shall encourage them to:

- *Organise* his/her mathematical work
- *Reason* mathematically
- *Control* several conditions (verification, adjustment, modification, reorganisation, awareness of contradictions, validation)
- *Choose/develop* efficient strategies/tools of problem solving
- *Reflect* on methods of mathematical work
- *Communicate* his/her results in a "mathematical" way (oral/written form, use of symbols, giving valid explanations)

Thus, we will be able to *identify* gifted children who:

- Ask spontaneously questions beyond given mathematical task
- Look for patterns and relationships
- Build links and mathematical structures
- Search for a key (essential) of the problem
- Produce original and deep ideas
- Keep a problem situation under control
- Pay attention to the details
- Develop efficient strategies
- Switch easily from one strategy to another, from one structure to another
- Think critically
- Persist in achieving goals

At the same time, we could nurture their curiosity, willingness to learn more about mathematics, provide them with an opportunity to go further in their mathematical learning, to create new structures, to pose new problems and thus *foster* the development of their mathematical abilities.

This approach is very demanding to the teaching. The teacher has to think constantly about challenging the students, look for different ways to stimulate children's work, demonstrate a high flexibility, ability to react spontaneously on changing conditions of the classroom situation, be ready to provoke students and to get provoked by students asking question which the teacher can not answer immediately.

A better understanding of how to help highly talented children to develop deeper mathematical thinking would lead to elaboration of efficient didactical approaches for all students. We shall agree with following general remark made by Young and Tyre (1992): "If we examine more closely what it is that makes prodigies, geniuses, gifted people, high achievers, champions and medallists, we may be better able to increase their number dramatically."

REFERENCES

Bachelard, G. (1938). *La formation de l'esprit scientifique*. Paris: Presses Universitaires de France.

Bednarz, N., & Poirier, L. (1987). *Les mathématiques et l'enfant (Le concept du nombre)- bande vidéo*. Montreal: UQAM.

Brousseau, G.(1997). *Theory of didactical situations in mathematics*. Dordrecht: Kluwer Academic Publishers.

Burjan, V.(1991). Mathematical giftedness—Some questions to be answered. In F. Moenks, M. Katzko, & H. Van Roxtel (Eds.), *Education of the gifted in Europe: Theoretical and research issues: Report of the educational research workshop held in Nijmegen (The Netherlands) 23–26 July 1991* (pp. 165–170). Amsterdam/Lisse: Swetz & Zeitlingen Pub. Service.

Dunham ,W.(1990). *Journey through genius*. New York: Penguin Books.

Greenes, C. (1981). Identifying the gifted student in mathematics. *Arithmetic Teacher*, 14–17.

Greenes, C. (1997). Honing the abilities of the mathematically promising. *Mathematics Teacher*, 582–586.

Kennard, R. (1998). Providing for mathematically able children in ordinary classrooms. *Gifted Education International, 13*(1), 28–33.

Krutetskii V. A.(1976). *The psychology of mathematical abilities in school children*. Chicago: The University of Chicago Press.

Kulm, G. (1990). New Directions for Mathematics Assessment. In G. Kulm (Ed.) *Assessing higher order thinking in mathematics*. Washington, DC: American Association for the Advancement of Science.

Lyons, M., & Lyons, R. (1989). *Défi mathématique. Manuel de l'élève. 3–4–5–6*. Laval: Mondia Editeurs Inc.

Lyons, M., & Lyons, R. (2001–2002). *Défi mathématique. Cahier de l'élève. 1–2–3–4*. Montreal: Chenelière McGraw-Hill.

Miller, R.(1990). Discovering mathematical talent. *ERIC Digest* #E482.

Mingus, T., & Grassl, R. (1999). What constitutes a nurturing environment for the growth of mathematically gifted students? *School Science and Mathematics, 99*(6), 286–293.

Programme de formation de l'école québecoise (2001). Québec, 2001.

Renzulli, J. (1977). *The enrichment triad model*. Connecticut: Creative Learning.

Ridge, L., & Renzulli, J. (1981). Teaching mathematics to the talented and gifted. In V. Glennon (Ed.), *The mathematical education of exceptional children and youth: An interdisciplinary approach* (pp. 191–266). Reston, VA: NCTM.

Shchedrovitskii, G. (1968) *Pedagogika i logika.* Unedited version (in Russian).

Sheffield, L.(1999) Serving the needs of the mathematically promising. In L. Sheffield (Ed.), *Developing mathematically promising students* (pp. 43–56). Reston, VA: NCTM.

Sierpinska, A.(1994). *Understanding in mathematics,* London: The Falmer Press.

Young, P., & Tyre C. (1992). *Gifted or able?: Realising children's potential.* Open University Press.

ACKNOWLEDGMENT

Reprint of Freiman, V. (2005). Problems to discover and to boost mathematical talent in early grades: A Challenging Situations Approach. *The Montana Mathematics Enthusiast, 3*(1), 51–75. Reprinted with permission from The Montana Mathematics Enthusiast. © 2005 Viktor Freiman

CHAPTER 8

MATHEMATICAL PROBLEM SOLVING PROCESSES OF THAI GIFTED STUDENTS

Supattra Pattivisan
The Institute for the Promotion of Teaching Science and Technology (IPST), Bangkok, Thailand

Margaret L. Niess
Oregon State University

ABSTRACT

The purpose of this study was to examine the problem solving processes of Thai gifted students as they solve non-routine mathematical problems. Five Thai gifted students participated in this study. Each student practiced the think aloud method before solving three non-routine problems focused on number theory, combinatorics, and geometry, respectively. Data sources included videotapes of the think aloud and the interview sessions, students' written solutions, and the researcher's field notes. The results generated a Thai model of problem solving process that detailed behaviors in each of four stages: understanding, planning, executing, and verifying. Metacognitive behaviors contributed to participants' activities in each stage. The findings also provided five categories of emerging evidence related to the students'

Creativity, Giftedness, and Talent Development in Mathematics, pages 185–207
Copyright © 2008 by Information Age Publishing
All rights of reproduction in any form reserved.

problem solving processes: advanced mathematical knowledge, willingness to consider multiple alternative solution methods, recollection and willingness to consider prior knowledge and experiences, reliance on affect, and parental and teacher support.

BACKGROUND OF THE STUDY

Problem solving is a form of inquiry learning where existing knowledge is applied to a new or unfamiliar situation in order to gain new knowledge (Killen, 1996; Sternberg, 1995). Engaging in problem solving requires coordinating multiple cognitive and metacognitive processes, selecting and deploying suitable strategies, and adjusting behavior to changing task demands (Montague, 1991). Various models are proposed to describe the processes that problem solvers use from the beginning until they finish their tasks. For instance, Polya's model consists of four stages: understanding the problem, devising a plan, carrying out the plan, and looking back (Polya, 1957). Later, Garofalo and Lester (1985) modified Polya's model to include cognitive and metacognitive components described in four stages: orientation, organization, execution, and verification. Montague and Applegate (1993) presents a model focused on seven cognitive processes (reading, paraphrasing, visualizing, hypothesizing, estimating, computing, and checking) and three metacognitive processes (self-instruction, self-questioning, and self-monitoring). Only two models (Garofalo and Lester's model; Montague and Applegate's model) that emphasize metacognitive processes have been used with gifted students in the literature (Garofalo, 1993; Montague, 1991; Montague & Applegate, 1993; Sriraman, 2003). Research on these models has indicated that gifted students use metacognitive strategies in their problem solving. The research reports substantial use of self-instruction throughout the problem, frequent self-questioning during and after reading the problem, and effective self-evaluation and self-monitoring activities (Montague, 1991; Montague & Applegate, 1993).

Mathematically-gifted students are identified as students who are able to do mathematics typically accomplished by older students. They are able to employ qualitatively different thinking processes in solving problems (Sowell, Zeigler, Bergwell, & Cartwright, 1990). At the same time, successful mathematical problem solving requires students to be able to select and use task-appropriate cognitive strategies for understanding, representing, and solving problems (Mayer, 1992; Schoenfeld, 1985). These abilities involve metacognitive knowledge, a knowledge that is necessary for higher-level learning and problem solving (Brown, 1978). Research studies have found that metacognitive knowledge and processes help problem solvers become more efficient at handling problems in three aspects: (a) defining the problem, and forming a mental representation of its elements, (b) selecting suitable plans and strategies for achieving a goal, and (c) identifying and mastering obstacles that facilitate progress (Davidson & Sternberg, 1998). Metacognition in problem solving involves the processes of planning, monitoring, and evaluating specific problems-particularly in making the mental representation and the selection of appropriate strategy (Flavell, 1992; McCormick, 2003). The use of metacognitive processes supports problem solvers during the solution process and improves their ability to obtain the goal. The more problem solvers control and monitor the strategies they use, the better are their abilities to solve a problem (Fortunanto, Hecht, Tittle, & Alvarez, 1991; Kapa, 1998; Swanson, 1990).

Non-routine problems are the type of problems where the students are not familiar with problem situations and they are not expected to have previously solved or have not met regularly in the curriculum. Non-routine problems demand thinking flexibility and extension of previous knowledge, may involve concepts and techniques which will be explicitly taught at a later stage, and may involve the discovery of connections among mathematical ideas (Schoenfeld, et al., 1999). Researchers have found that more difficult problems have the potential to activate metacognitive functioning to the extent that good problem solvers consciously regulate and control their cognitive processes (Montague & Applegate, 1993). Additionally, gifted students prefer to solve non-routine problems because of the challenge of working with these problems (Garofalo, 1993). Thus, non-routine problems are more likely to activate gifted students to demonstrate their high ability in problem solving.

Researchers have studied how secondary gifted students solve non-routine mathematics problems (Garofalo, 1993; Lawson & Chinnappan, 1994; Montague, 1991; Sriraman, 2003). Results indicated that gifted students spend much time rereading and translating the problems into their own words (Garofalo, 1993; Montague, 1991; Sriraman, 2003). This paraphrasing ability supports them in understanding the problem and indicates one way they differ from other students in problem solving. They are more verbal than other students and their verbalization increases when they are confronted more difficult problems (Sriraman, 2003). They recall theorems for generating new information, apply prior knowledge in the problem and use it to access further relevant knowledge (Lawson & Chinnappan, 1994; Sriraman, 2003). Gifted students identify their assumptions in the problem, frequently set up an equation or algorithm after reading, and generally divide the problem into sub-problems. They identify a goal before developing their solution plans. They solve the problems systematically, and use efficient strategies. They redo the problems by working through the whole problems, rereading them, redoing computations, and checking steps and processes (Montague, 1991; Sriraman, 2003).

However, the participants in the research literature are from a western culture, a culture distinct from Asian cultures, including the Thai culture. Asian students receive substantially more exposure to mathematics at school and at home than U. S. students (Geary, 1996). Asian cultures emphasize and give priority to mathematical learning (Hatano, 1990). Social contexts such as culture and language structure play a role in problem solving ability, especially in recognition, definition, and representation of a problem. They also suggested that the social context may facilitate understanding the problem and thinking divergently about the solution (Pretz, Naples, & Sternberg, 2003). It is logical to assume that cultural differences may lead students to perform in different ways when engaged in mathematical problem solving.

Considering the current status of gifted education in Thailand, not enough knowledge and resources exist for gifted education for nurturing the promotion of gifted abilities. These problems result in an inefficient system and are unsuccessful in nurturing gifted students (Office of the National Education Commission [ONEC], 2004). Not much research has been done about mathematical problem solving of Thai gifted students. Only two related studies focused on the development of an enrichment program, rather than understanding the problem solving process of gifted students (Klaimongkol, 2002; Thipatdee, 1996). Consequently, there is a need to examine how secondary Thai gifted students think and what strategies they employ when solving non-routine problems. The purpose of this study was to examine two research questions:

1. What is the nature of the problem solving processes that Thai gifted students use as they engage in solving non-routine mathematical problems?
2. What metacognitive behaviors do Thai gifted students exhibit when engaged in mathematical problem solving?

RESEARCH SETTING

In Thailand, one of the ways to promote abilities of gifted students' abilities in mathematics is through participation in the International Mathematical Olympiad (IMO). The IMO is the annual world championship mathematics competition for secondary students. This event is recognized as a strategy of fostering mathematical talent (Wieczerkowski, Cropley, & Prado, 2000). The Thai government has given the Institute for the Promotion of Teaching in Science and Technology (IPST) the responsibility of selecting the Thai representatives for this annual contest since 1989. For the Thai Mathematical Olympiad (TMO) project, secondary gifted students from around the country complete in two rounds of examinations for entrance into the project each year. Both examinations focus on non-routine mathematical problems with the mathematical content limited to grade 11 according to the national curriculum. In June 2005, a total of 7,982 gifted students applied for the TMO project. Among these students, 6,861 students took the first round examination which examined their mathematical problem solving abilities through multiple-choice questions and short answer questions. Only 42 students were selected from the first round exam to take the second round examination, using all open-ended mathematics questions where students are required to show their written solutions. Finally, 24 students were selected from the second examination to participate in the TMO project. These 24 students enrolled in the TMO training camp at IPST. During the training, examinations are given and used to select six mathematically-gifted students as the Thai representatives for the IMO. This research study was conducted while 24 gifted students were participating in the TMO training camp at the IPST.

PARTICIPANTS

The researcher used purposeful sampling to select participants out of the pool of 24 gifted students within the camp. Although these students were selected by the entrance examinations where the mathematical content was limited to grade 11, their grade levels varied from grades 8 to 12. Thus, the students had different levels of mathematical background. However, they have had prior experiences in solving non-routine mathematical problems from the entrance examinations that also required them to provide written solutions. With the definition of gifted students used in this study, student participation was limited to students in grades 8–10 even though these students were able to do mathematics typically accomplished by older students (Sowell, et al., 1990). Other criteria were also used to identify students with similar mathematics background. Students were considered if they (a) had similar scores on the second round of the entrance examination to the TMO project, (b) did not participate in the training camp in the previous year, and (c) were in grade 10 or below. Five of the seven gifted students who matched the selection criteria were selected to assure a diversity of school, grade, gender, and age. The participating students consisted of four males and one female. Their ages at the time of participation ranged from 13 years 3 months to 16 years 6 months. They were studying in grades 8, 9, and 10 of four different schools. Three schools were in Bangkok, the capital of Thailand, and another school was outside Bangkok. All participants were assigned pseudonyms as Pradya, Sira, Wude, Nipa, and Sakda for reporting results.

PROBLEM SELECTION

The mathematical problems for this study consisted of three non-routine problems that were selected and modified from a variety of sources, including mathematical journals, textbooks and examination contests (ApSimon, 1991; Covington, 2005; Gardiner, 1987; Krantz, 1996; Posamenteir & Schulz, 1996; Posamenteir & Salkind, 1996; Schoenfeld, 1985). The researcher also considered the training curriculum before developing a pool of problems that met the following criteria:

- The problems are non-routine; that is, the students are not familiar with the problem situations, they are not expected to have solved the problems previously, nor have they worked on them in the mathematics curriculum. Non-routine problems demand thinking flexibility and extension of previous knowledge, may involve concepts and techniques that will be explicitly taught at a later stage, and may involve discovery of connections among mathematical ideas (Schoenfeld, et al., 1999).
- The problems cover mathematical content areas in number theory, combinatorics, and geometry, which are the major subjects in training students for the IMO.
- The problems challenge students' thought processes; that is, the problems must examine the upper levels of the cognitive domain according to Bloom's Taxonomy: analysis, synthesis, and evaluation (Bloom, et al., 1956).
- The solutions do not require mathematical concepts and skills that students have not learned according to the national curriculum.

A pool of problems consisted of 13 problems: 6 number theory problems, 3 combinatorics problems, and 4 geometry problems. The pool was examined for content validity by seven experts with expertise in mathematics and mathematics education. The experts selected one problem in each area and suggested important feedback for the study. The three problems used in the study are presented in Figure 8.1.

- **Problem One**: Does a Friday the 13th occur every year? Explain your reasons.

- **Problem Two**: In a tournament, there are 15 teams. Each team plays with every other team exactly once. A team gets 3 points for a win, 2 points for a draw, and 1 point for a loss. When the tournament finishes, every team received a different total score. The team with the lowest total score is 21 points. Explain why the highest total score team has at least one draw.

- **Problem Three**: Let *ABC* be an isosceles triangle with *AB* = *BC*. Angle *ABC* equals 20 degrees. Point *D* is on *AB* such that angle *ACD* equals 60 degrees. Point *E* is on *BC* such that angle *EAC* is equal to 50 degrees. Find the value of angle *CDE*.

Figure 8.1   Three problems used in the study.

DATA COLLECTION

Data collection was in a one-to-one setting between the participant and the researcher. Each participant made three appointments for solving one out of three problems using the think aloud method followed by an individual interview. The appointments were weekly during the training and were conducted before or after the training time. For the first appointment, the researcher described an overview of procedures for the participant to follow including the think aloud protocol. Each student read through the instructions and asked any questions before beginning to work on the sample problem. Practice with the think aloud method took about 15 minutes. The researcher replayed the videotape and discussed strategies in improving the participant's skills for mastering this technique.

Next, the participants received Problem One, began reading the problem aloud and asked any questions to make sure they understood the wording in the problems before solving the problem. Not all students asked questions at this point. Participants spoke aloud describing their thinking while writing their solutions on the paper. They were allowed to use as much time as they needed in solving each problem. On average, the participants took about 20 minutes per problem, followed by a 15-minute interview. Each interview was intended to elicit the participants' backgrounds and their thought processes in solving the given problem. At the end of each appointment, the researcher confirmed that the participant was not to disclose the nature of the problems with the other participants.

RESULTS

Transcribed data were analyzed to conceptualize a model of the Thai students' problem solving process. Two models (Garofalo & Lester's model (1985) and Montague & Applegate's model (1993)) were used as references for analyzing the metacognitive aspects since both reference models focused on metacognitive behaviors and have been used to describe gifted students' problem solving processes in the literature. Because classifying behaviors as exactly cognitive or metacognitive processes is difficult, some researchers have used the term cognitive-metacognitive (Garofalo & Lester, 1985; Artzt & Armur-Thomas, 1992; Pugalee, 2001). For this study, these behaviors were correlated to the problem solving stages and were assigned as metacognitive processes. Overall, the gifted students in this study have successfully demonstrated their processes in solving the non-routine mathematical problems. Typically, the students solved Problem One and Problem Two without any hesitation although some did not completely solve Problem One (i.e., they forgot to consider leap years in their solutions). Although, at first, two students had some difficulty searching for a solution for Problem Three, all the students eventually succeeded.

THAI STUDENTS' PROBLEM SOLVING MODEL

The Thai students' behaviors discussed were conceptualized into a four-stage model as illustrated in Figure 8.2. These stage names, labeled as understanding, planning, executing, and verifying, were modified from the study of Garofalo and Lester (1985). Understanding was the first important stage in guiding the gifted students toward success in solving the problem set. After the students read the problems aloud, they identified the questions. The given information was stated, interpreted, and represented with pictures or tables as well as organized into a systematic format. Rereading the problem was used for checking the correctness of their representations. Students' prior knowledge was necessary when they interpreted the given information and referred to any relevant concepts before developing a solution plan. Reflections on problem difficulty and familiarity were also established in this stage.

The second stage was *planning*. Students generated new information and represented the problems with pictures, symbols, or tables as well as organized them into a plan. Efficient strategies, such as drawing pictures, making tables, or looking for patterns were employed with the application of relevant mathematical concepts in number theory, basic of counting and geometry in solving the problems. The plans were reevaluated and determined whether they were valid. New plans were derived if the current plans were determined to be invalid.

The third stage was *executing*. The students proposed a final answer by carrying out their computations in this stage. The students made logical mathematical statements to support their plans and finally stated their conclusions.

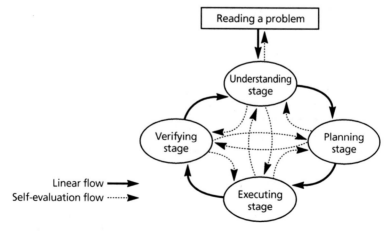

Figure 8.2 Thai students' problem solving model.

The last stage, *verifying*, involved the students checking their written solutions. During this stage, they may have also reread the problem to assure their solution.

As students were engaged in solving the problems, their thinking processes did not proceed in a linear order from the *understanding* stage to the *verifying* stage. Their efforts showed that while they performed each of the stages, they did not proceed each stage linearly from the first to the last one. Thus, an important found in this study was that this model is not linear. In each stage, the term *self-evaluation* from Montague and Applegate (1993) was used to label when the problem solvers monitored their thinking and efforts and demonstrated affective behaviors as they worked in solving the problem. In other words, self-evaluation affected the participants' actions in each of the stages of the model. Figure 8.3 provides a more complete description of each stage in the model.

Stage: Understanding
(1) Identify the problem
  • Read/reread/restate the problem, the given information, and the question
(2) Analyze the problem
  • Represent the problem with pictures or tables
  • Clarify/interpret/organize the given information
  • Connect with prior experience
  • Reflect on the problem
(3) Self-evaluation

Stage: Planning
(1) Devise a plan
  • Manipulate the given and generating information
(2) Assess a plan
  • Apply prior knowledge/ mathematical concepts/ theorems
  • Use strategies (look for pattern/make a table)
  • Predict possible answers/use estimation
(3) Revise a plan
  • Determine the plan makes sense
  • Change the plan if it is not working
(4) Self-evaluation

Stage: Executing
  • Carry out computations
  • Make logical mathematical statements
  • State the conclusion/the answer
  • Self-evaluation

Stage: Verifying
  • Check results for reasonableness
  • Reread the problem and solutions for checking
  • Move to a new plan based on verifying results
  • Self-evaluation

Figure 8.3   Thai students' problem solving stage description.

DESCRIPTION OF THE THAI STUDENTS'
PROBLEM SOLVING PROCESSES

Understanding. During this stage, the students demonstrated specific behaviors in trying to understand the problem after reading it aloud. Their actions included reading the problem, stating and/or restating the given information, stating the question using some words in the question, and restating the question. Since Problem One was in the form of a question and was very short, they restated the question in their own words or stated the question. When the participants were faced with more given information in the longer problems, they tried to state the given information from their understanding. In Problem Two, students stated the given information part-by-part at different times. Their actions showed that the students did not consider all of the given information as a whole at a single time. Students also thought about what the problem was asking and why it was asked.

> **Wude:** What does the problem ask? It asks about the highest total score.
>
> **Sakda:** Why does the lowest total score team gets 21 points? Why? And, why the highest score team has at least one draw?

For a geometry problem like Problem Three, the students stated all the given information while drawing a picture in order to make sure that they had everything that the problem provided before continuing with their solution. At the same time, the figures students drew indicated that they used their representations in their solution processes. A picture representation was important not only for understanding a geometry problem but also helped them to devise a plan for the solution. Some students drew more than one picture; they drew a new picture repeatedly when they could not figure out a way to solve the problem. They believed that a big picture provide them better visualization. The evidence showed other representations that students used in their processes to help them understand the problem, such as making a table or writing a calendar.

> **Nipa:** I am thinking about the calendar. There are 7 days in a calendar. I am writing a calendar now.
>
> **Pradya:** Make a year table, first.

After students stated the given information, they tried to clarify the given information as much as they could; in fact their explanations were based on their prior knowledge in mathematics. As in Problem Two, all of them carefully considered the sentence "Each team plays with every other team exactly once." Their interpretations indicated that they had a concept in basic of counting.

> **Pradya:** Okay, each team plays 14 times.
> **Sira:** Amm…15 teams play with every other team one time. It means the first team meets the other teams 14 times.
> **Nipa:** Each team plays 14 times.
> **Wude:** Let's see team 1. Team1 must meet the other teams 14 times.
> **Sakda:** That is, a team meets other teams 14 times. A team plays 14 times.

In Problem Three, the students stated all the given information part-by-part while drawing a picture. At the same time, they integrated their prior knowledge about an isosceles triangle with the given information and calculations to find other angles and length of the sides while drawing.

During the understanding stage, the students reflected on the problems in terms of their familiarity with the problem and their difficulty with it as they worked to understand in Problem One and Problem Three.

> **Sakda:** For this problem, I had never seen before. Friday the 13th.
> **Sira:** How do we solve this? I'm confused. Friday the 13th.
> **Nipa:** I cannot solve it.

Planning. During the planning stage, the Thai students searched for solution plans by selecting given information and generating new information. When they devised their plans for Problem Two, they focused on a particular sentence: "A team gets 3 points for a win, 2 points for a draw, and 1 point for a loss." This statement helped them generate the same information; each game produced 4 points.

The next given information (When the tournament finishes, every team receives a different total score. The team with the lowest total score is 21 points.) was considered and provided the constraints for their plans. Manipulating between the given and new information led students to identify the following plans to find the total games played and how many scores the highest team received before proving what was asked in the question (i.e., why the maximum score team must have at least one draw).

Students described their reasons behind their plans for Problem One as follows. Wude's plan started by setting January 1st as Friday and then he looked for a Friday the 13th in the same year. He continued his search by moving January 1st to Saturday, Sunday, and so on until he identified 7 cases for the year (years with 28 days in February), and then similarly for the other 7 cases for years with 29 days in February. At first, Nipa planned to solve this problem using a contradiction as she said, "How do I solve it? I think a Friday the 13th occurs every year. Or I assume that a Friday the 13th does not occur every year. Then, I find a contradiction." While she did not end up using this method, she considered various types of mathematics problems to identify a way to solve this problem as she said "Is this problem a combinatorics problem or a number theory problem? Now, I am finding the way to solve it."

Some students had difficulty developing plans for Problem Three. At the same time, the evidence indicated that they searched for a variety ways to solve this problem. Sakda tried to solve it with both Euclidean geometry and trigonometry. He searched for several ways to solve the problem, such as extending a line, drawing a circle, and using the Law of Sines. He also recalled what he had learned the previous day when finding a solution plan that used a compass to draw a circle that passed through points G and D and met BC at the point F. Nipa also had problem difficulty as she considered many plans for Problem Three. In her plans, she drew both a perpendicular line and a parallel line, but those lines did not work. Finally, she found a way to solve the problem as she drew a line CF and found an answer.

Students predicted the answer when they set up the plan for Problem One. For instance, Nipa, Sira and Pradya guessed that Friday the 13th occurred each year and Sakda knew from his prior experience that it occurred every year.

When the students developed the plans, they referred to mathematical concepts that they considered might be used in their plan. For Problem Two, Sakda first considered the Pigeonhole Principle, but he did not end up using it in the solution plan. Wude mentioned modulo 7 for Problem One, and trigonometry for Problem Three. He then used both concepts in his solution plans. These techniques indicated that the students used their knowledge of mathematics in their solutions.

> **Sakda:** Or we need to use Pigeonhole Principle.
> **Wude:** I consider the date with modulo 7.
> **Wude:** I'm thinking about how to use trigonometry to solve this.

When assessing a plan, the students applied their relevant prior knowledge in number theory, basic of counting, and geometry. For Problem One, they used the concept of remainders to show that a Friday the 13th occurred every year. Wude initially set January 1st as Friday and then looked for other Friday the 13ths in the same year. He continued this process by moving January 1st to Saturday, Sunday, and so on until he got 7 cases for the year that has 28 days in February, and the other 7 cases for 29 days.

In Problem Two, the students had a goal to prove why the highest total score team had at least one draw. Although they proved this result with a method of proof by contradiction, their reasons varied. Pradya considered that the highest total score team played 14 times and got 35 points. He reasoned that if this team did not get any draw, the team could only get wins and losses. Thus, the score needed to be an odd number because there were 3 points for win or 1 point for loss. When an even number was multiplied by an odd number, he recognized the result was an even number. Since 14 is an even number, the total score needed to be an even number too. But, 35 was an odd number.

Sira thought that if this team did not get any draw, the team must win 11 times and lose 3 times resulting in the scores 33 + 3 = 36 points. This score meant the highest score team with no draws must get the score of at least 36 points, but this team only received 35 points. Nipa supposed that the team did not have any draws. For her, the equation that represented the scores when this team played 14 times and won x times would be $3x + (14 - x) = 14 + 2x = 2(7 + x)$ points. She noticed that $2(7 + x)$ was an even number, but 35 was an odd number.

Wude considered the value of x and y to represent the number of wins and draws. After solving equations, he got $y = 21 - 2x$. He also knew that $21 - 2x$ was an odd number and x was more than zero. So, he concluded that $y \geq 1$. Sakda considered that if this team had only wins, its score would be 42 points. Each time when the team got a loss, its score decreased 2 points from 42 points. But, its score was 35 points. This result was impossible if the team was to get 35 points as it decreased 2 by 2 from 42.

In Problem Three, Wude applied the Law of Sines in his plan. With this application, he found the trigonometry equations and tried to find the value of angle *CDE*. However, he did not solve these equations. Rather, he used trial and error as he substituted the value for the angle in the equation until he found the answer.

During the planning stage, the students used efficient strategies, such as making tables, using symbols, and looking for patterns to represent the information. In Problem One, Pradya assumed the day for January 13th and used a variable x to represent it. The variable x represented Sunday, Monday, Tuesday, Wednesday, Thursday, Friday, or Saturday. He also put the remainders that he found in a table. Sakda made a table that consisted of months, number of days in that month, and remainders for solving Problem One.

Three students (Sira, Wude, and Nipa) used symbols when they set equations for Problem Two. The students also used variables to represent the value of the angle they were trying to find in Problem Three. Then they started comparing this angle with the others. Students looked for patterns when they solved Problem One. For example, Wude looked for a pattern of a Friday the 13th after he set January 1st to a Friday and found the other Fridays in January by counting day by day. He showed all 14 cases that January 1st fell on Friday, Saturday, Sunday, ... Thursday for years with Februarys that have 28 days and those with 29 days.

There was evidence that the students verified that their plans made sense, that they looked for efficient plans, and that they changed their plans during this stage.

> **Sakda:** Only 14 cases. Is it good to write a calendar? Not good.
> **Wude:** Are there any other ways that easier than this?"
> **Pradya:** Which lines do I need to extend? The extended line, which is the most useful to get the answer? Line. Line. Which is the best line? Okay, this line may be good.

Executing. During this stage, the Thai students carried out computations with the application of mathematical formulas to achieve their goal to find the final answer. As in Problem Two, three students stated "the combination of 15 and 2" when they calculated how many games were played in this tournament. This statement indicated that they knew the binomial coefficient formula $C_{(15,\,2)}$. Nipa did not mention these words, but she calculated the result in the same way. Sira calculated this result by finding the sum of a sequence as he said "the total number of games play is $= 14 + 13 + 12 + \ldots + 3 + 2 + 1 = (14 \times 15)/2 = 105$ times." All students also used the sum of arithmetic sequence formula for calculating the sum of a sequence 21, 22, 23, ..., 35 in solving this problem. Some students solved equations to find an unknown variable as Wude and Sira did in Problem Two.

> **Wude:** From these two equations, we get $2x + y = 21$.
> **Sira:** Subtract the second equation from the first equation; we get $2a + b \geq 21$. Oh ... wait a minute if we suppose the highest score team did not have any draw, that is, $b = 0$. Then we get $3a + c \geq 35$ and $a + c = 14$. Since, $2a \geq 21$ then $a \geq 10.5$.

Students usually made logical mathematical statements that supported their plans before stating the conclusions or the final answer.

> **Nipa:** It occurs every year because when I suppose January 13th is Friday, I found that the 13th of the other months fall on the rest six days. Therefore, there is a Friday the 13th in every year.
> **Sira:** If this team did not have any draw, it must win 11 times and loss 3 times and gets the scores $33 + 3 = 36$ points.

This means the highest score team with no draw must get the score at least 36 points, but it get only 35 points, contradiction. Therefore, the highest score team has at least one draw.

Verifying. During the final stage, *verifying*, the Thai students sometimes checked their processes and results to make certain that the solution made sense. They usually revised the solution plans when the plans did not work. The students rechecked what was done and were able to explain reasons for their solutions. When they verified the solutions, they reread the problem and examined all their written responses in cyclical processes as they attempted to verify local plan as to its usefulness for solving the problems. Only one student, Nipa showed that she paid attention to verifying the solution for Problem Two. She stopped speaking during this stage for 3–4 minutes. So, the researcher asked her to speak aloud what she was thinking.

Researcher: What are you thinking?
Nipa: I am checking my solution. I examine what I wrote. It is correct or not. I return to read the question and read my solution again. I am done.

Although the other students did not directly demonstrate checking their solutions, obviously Nipa had. The students did examine what was done in the plan before confirming their answers.

Wude: I have already checked all cases, 14 cases. A Friday the 13th occurs in all cases. Therefore, it occurs every year.
Pradya: We got them all. We have all 12 months in a year.

Wude also checked his computations again after he got the result that used the binomial coefficient formula to find the total numbers of games played. Then he restated the result to confirm it was correct.

Wude: Is it correct? Correct. It is the total number of games, 105 times.

After Sakda carried out his calculations to find how many games were played, he noticed that the result was too much as he said "It is so much. Did I do something wrong? There are 105 games of 15 teams." At this point, he checked to make certain that the answer made sense in his mind. However, he did not specifically show a way to verify the solution.

Self-Evaluation. In addition to each of the above stages, students frequently exhibited self-evaluative statements that helped them continue working

TABLE 8.1   Examples of Excerpts for Self-Evaluation

Self-Evaluation	Examples of Excerpts
Self-monitoring	What's next? What am I doing now?
	Sakda: Where is the way to solve? How to find the answer?
	Sakda: Keep trying. Keep going.
Affect	**Nipa:** Ohh…I see. I got it.
• Confidence	**Wude:** Is it correct?
	Pradya: I do not know I can solve it or not.
	Pradya: From what I did, I cannot find anything wrong.
	Sakda: I think I can figure it out.
Affect	**Nipa:** I don't know what I'm going to do. I cannot solve it.
• Difficulty/Frustration	**Sira:** How do we solve this? I'm confused.
	Sakda: Ohh…I cannot think about wording to explain.
	Nipa: Draw a new picture. This picture is too small. It makes me so confuse. I cannot think when I see it.
	Sakda: I cannot figure it out. I'm so confused. I am very stressed.
	Sakda: Why I cannot think about them? I cannot find anything.
Affect	**Wude:** We must think slowly. Think slowly.
• Effort	**Wude:** Amm… it needs to think a lot.
	Sakda: Ohh…I am lazy to think about the sum.
	Pradya: I am very lazy now.

on the problem until they had finished the tasks. The term "self-evaluation" was used as a category for coding in the study of Montague and Applegate (1993) when the students evaluated themselves as problem solvers or used "I" statements about their performance. For this study, self-evaluative statements were divided into two types. First, students demonstrated self-monitoring as they monitored their work on the problems until they got complete solutions. Second, the students demonstrated affect statements as they evaluated themselves as problem solvers in terms of how much confidence they had, their difficulties and frustrations, and their efforts while solving the problems. Examples of self-evaluation statements in each of these categories are provided in Table 8.1.

DISCUSSION OF FINDINGS

This study described a model that captured the solution processes of the Thai gifted students. The results indicated that students' solution processes were based on logical analysis and systematic strategies. They showed high ability in verbalizing and explaining their thoughts and reasoning for their solutions. This ability demonstrated their understanding of mathematical structure and strategies, similar to those described in Heinze's study (2003). The study findings were consistent with other studies in the literature, where participants were from Western cultures. In particular, the results were consistent within the context that gifted students accessed problem representation strategies, such as drawing pictures, making tables, or looking for patterns in order to facilitate their understanding (Gorodetsky & Klavir, 2003; Montague, 1991; Montague & Applegate, 2000; Sriraman, 2003). The Thai gifted students and the students in the Western culture research applied their prior knowledge to a problem or to an unfamiliar situation; they made use of a variety of mathematical knowledge, recalling and relying on theorems to generate additional relevant information (Gorodetsky & Klavir, 2003; Lawson & Chinnappan, 1994; Overtoom-Corsmit, Dekker & Span, 1990). The Thai gifted students were observed to increase their conversations as they were confronted with more difficult problems; this result was similar to the actions of the American students in Sriraman's study (2003).

Based on an analysis of the results, major emerging evidence related to the students' problem solving processes consists of five categories: advanced mathematical knowledge, willingness to consider multiple alternative solution methods, recollection and willingness to consider prior knowledge and experiences, reliance on affect, and parental and teacher support. First, advanced mathematical concepts were integrated in students' solutions such as the Law of Sines, the binomial coefficient formula, and the sum of arithmetic sequence formula. All these concepts were usually taught in grades 11–12 of the Thai national curriculum. Yet, these students who were not yet in those grades had in some way gained this prior knowledge. This result indicated that these participants had conceptual understandings in mathematical concepts in higher levels, a characteristic of gifted students. The result also suggested a high ability in problem solving for applying these advanced concepts toward the completion of a correct solution.

Second, the students searched for alternative solution methods as they worked on the problems. These students tried to understand and worked with the problem using a variety of approaches when they had difficulties. As in Problem Three, Sakda and Nipa demonstrated their willingness and abilities to consider different paths in their solution rather than insisting on the unproductive paths. However, the ways they searched for alternative paths depended upon their prior experiences and beliefs about working with mathematical problems.

Third, the influence of the students' prior knowledge was evident. Students applied techniques and strategies that they had used previously. Wude reported that he usually used a trigonometry approach in a geometry problem because, from his experience in solving this type of problem, more than 60% of the time this method led him to the solution. He also feared that he could not solve the problem if he used Euclidean geometry.

However, he planned to try drawing some lines if he could not find the answer by using trigonometry. For Problem One, Sakda had experienced finding a Friday the 13th in the calendar in every year as he said "Actually, I always look for a Friday the 13th in new calendars every year. There exists Friday the 13th in every calendar year. For example, it will be in January next year, Friday January the13th. I have found it to occur every year. This is a math problem. Friday the 13th should relate to number theory. I think there is another problem I have read. That problem asked about the New Year day."

Fourth, affective behaviors (behaviors relying on emotion more than cognition, thought and action) played an important role in the problem solving process of all five gifted students. This result was consistent with the findings of Carlson and Bloom (2005) and DeBellis (1998). As Goldin (2000) described, the ways that affect is or can be utilized by mathematical problem solvers to guide their steps and to their influence cognition in a constructive manner increased their problem solving power. In this study, affect was evident in terms of self-confidence, frustration, and effort. The students relied on their confidence to monitor their frustration and anxiety, turning these feelings into motivation that ultimately led them to success in finding a solution. Their motivation maintained their interest, encouraging them to continue working efficiently on the problems. Overall, the students expressed positive emotions while they attempted to solve the problems. One student indicated that he was stressed due to the level of difficulty in one of the problems, even though he was not under a time constraint as Sakda said "I felt stress. However, this is not an examination, and I can use as much time as I need. So, I just kept going."

Fifth, parental and teacher support were clearly noted as Wude responded to the interview question. He believed that their strong support helped him in becoming a good problem solver.

> **Researcher:** What makes someone a good problem solver in your opinion?
> **Wude:** Everything, I think. For me, I have excellent support from my parents. My teachers also helped me a lot. My mother always looked out for any good math problems out there for me to solve. I also really like to read math textbooks.

This student's response provided not only an example of teacher and parental involvements, but also illustrated Thai cultural influences. This evidence is consistent with the literature; in Asian cultures, parents consider their children's education their highest priority in their upbringing. Their children's education is emphasized and priority is given to mathematical learning (Hatano, 1990; Geary, 1996). This cultural value in mathematics also reflected differences in the investment of children, parents, and teachers in learning mathematics (Geary, 1994; Stevenson & Stigler, 1992). Perhaps the focus of the Thai culture on importance of children's education is an important factor in these students success in mathematical problem solving.

LIMITATIONS

Although the students had never seen the problems previously, there was a possibility that the results may be biased by the students' prior knowledge. Their prior mathematical background, knowledge, and problem solving experiences might have affected how they approached the problems and their solutions. The number of problems presented to the students was small and limited to some content areas in mathematics. The specific types of problem may have influenced their performances if they were not comfortable or did not specialize in these areas of content. The participants did not have any difficulty verbalizing their thought process, even though it was their first experience exercising the think aloud method. It is possible that the think aloud method may have improved their thinking processes and better assisted their approaches compared to the conventional thinking method. The participants may not have reported all of their thoughts because they had to spend extra effort applying the think aloud technique. In addition, an assumption was made that the students had answered all interview questions honestly without bias or concern for self-esteem.

IMPLICATIONS FOR FUTURE RESEARCH

The study was conducted while five participants were in the training camp. Results may be different for a larger number of participants extending over a longer period of time. Future study is needed to observe and compare the processes and behaviors when students are treated in their school environment, rather than in a training camp experience. Having no time constraint helped decrease students' pressure and supported them in activating their problem solving ability. Future research is needed to consider time as a factor when students work on non-routine mathematical problem in order to promote students searching for their own solution path. Results indicated that the Thai gifted students applied prior knowledge effectively, used a variety of knowledge, and had a high ability in verbalizing and explaining their reasons for their solutions. They are also capable of integrating advanced mathematical concepts in their problem solving. Thus, future research must recognize these factors and examine how they affect the process and ability in problem solving of gifted students. The findings indicated that gifted students employed self-evaluation statements during the think aloud session to help them be successful problem solvers. It would be reasonable to explore other variables related to these processes and ways to improve them for more efficient work during problem solving.

REFERENCES

Alexander, J. M., Carr, M., & Schwanenflugel, P. J. (1995). Development of metacognition in gifted children: Direction for future research. *Developmental Review, 15,* 1–37.

ApSimon, H. (1991). *Mathematical byways in ayling, beeling, and ceiling.* Oxford, NY: Oxford University Press.

Artzt, A. F., & Armour-Thomas, E. (1992). Development of a cognitive-metacognitive framework for protocol analysis of mathematical problem solving in small group. *Cognition and Instruction, 9,* 137–175.

Bloom, B., Englehart, M. Furst, E., Hill, W., & Krathwohl, D. (1956). *Taxonomy of educational objectives: The classification of educational goals. Handbook I: Cognitive domain.* New York: Longman.

Brown, A. (1978). Knowing when, where and how to remember: A problem of metacognition. In R. Glaser (Ed.), *Advances in instructional psychology* (pp. 77–165). Hillsdale, NJ: Lawrence Erlbaum Associates.

Carlson, M. P., & Bloom, I. (2005). The cyclic nature of problem solving: An emergent multidimensional problem solving framework. *Educational Studies in Mathematics, 58,* 45–75.

Covington, J. (2005). Solutions to January calendar. *Mathematics Teacher, 98,* 334–336.

Davidson, J. E., & Sternberg, R. J. (1998). Smart problem solving: How metacognition helps. In D. J. Hacker, J. Dunlosky, & A. C. Graesser (Eds.), *Metacognition in educational theory and practice* (pp. 47–68). Mahwah, NJ: Erlbaum.

DeBellis, V. A. (1998). Mathematical intimacy: Local affect in powerful problem solvers. *Proceedings of the 20th annual meeting of the North American Group for the Psychology of Mathematics Education* (pp. 435–440). Columbus, OH: ERIC Clearinghouse for Science, Mathematics, and Environmental Education.

Flavell, J. H. (1992). Metacognitive and cognitive monitoring: A new area of cognitive development inquiry. In T. O. Nelson (Ed.), *Metacognition-core readings,* (pp. 3–8). Library of Congress.

Fortunanto, I., Hecht, D., Tittle, C. K., & Alvarez, L. (1991). Metacognition and problem solving. *Arithmetic Teacher, 39,* 38–40.

Gardiner, A. (1987). *Mathematical puzzling.* Oxford, England: Oxford University Press.

Garofalo, J. (1992). Number-consideration strategies students use to solve word problems. *Focus on Learning Problem in Mathematics, 14,* 37–50.

Garofalo, J. (1993). Mathematical problem preferences of meaning-oriented and number-oriented problem solvers. *Journal for the Education of the Gifted, 17,* 26–40.

Garofalo, J., & Lester, F. K. (1985). Metacognition, cognitive monitoring, and mathematical performance. *Journal for Research in Mathematics Education, 16*, 163–176.

Geary, D. C. (1994). *Children's mathematical development: Research and practical applications.* Washington, DC: American Psychological Association.

Geary, D. C. (1996). Biology, culture, and cross-national differences in mathematical ability. In R. J. Sternberg, & T. Ben-Zeev (Eds.), *The nature of mathematical thinking,* (pp. 145–171). Mahwah, NJ: Erlbaum.

Goldin, G. A. (2000). Affective pathways and representation in mathematical problem solving. *Mathematical Thinking and Learning, 23,* 209–219.

Gorodetsky, M., & Klavir, R. (2003). What can we learn from how gifted/ average pupils describe their processes of problem solving? *Learning and Instruction, 13,* 305–325.

Hannah, C. (1990, April). *Metacognitive strategies used by learning-disabled gifted students.* Paper presented at the annual meeting of the American Association for Educational research Association, Boston, MA.

Hatano, G. (1990). Toward the cultural psychology of mathematical cognition. *Monograph of the Society for Research in Child Development, 55,* 108–115.

Heinze, A. (2003, July). *Mathematically gifted elementary students' problem solving strategies: Significant differences to "non-gifted" students.* Paper presented at the biennial conference of the World Council for Gifted and Talented Children. Adelaide, Australia.

Kapa, E. (1998). A metacognitive support during the process of problem solving in a computerized environment. *Educational Studies in Mathematics, 29,* 317–336.

Killen, R. (1996). *Effective teaching strategies: Lesson from research and practice,* Sydney, Australia: Social Science Press.

Klaimongkol, Y. (2002). *The development of an instructional process by applying a problem-based learning approach to enhance mathematical competencies of prathom suksa five gifted students in mathematics.* Unpublished Doctoral Dissertation, Chulalongkorn University, Bangkok, Thailand.

Krantz, S. G. (1996). *Techniques of problem solving.* Providence, RI: American Mathematical Society.

Lawson, M. J., & Chinnappan, M. (1994). Generative activity during geometry problem solving: Comparison of the performance of high-achieving and low-achieving high school student. *Cognition and Instruction, 12,* 61–93.

Mayer, R. E. (1992). *Thinking, problem solving, cognition.* New York: Freeman.

McCormick, B. C. (2003). Metacognition and learning. In W. M. Reynolds, & G. E. Miller (Eds.), *Handbook of Psychology,* (pp. 79–102). New York: John Wiley & Sons, Inc.

Merriam, S. B. (1998). *Qualitative research and case study applications in education.*San Francisco, CA: Jossey-Bass Publishers.

Montague, M. (1991). Gifted and learning disabled gifted students' knowledge and use of mathematical problem-solving strategies. *Journal for the Education of the Gifted, 14,* 393–411.

Montague, M., & Applegate, B. (1993). Middle school students' mathematical problem solving: An analysis of think-aloud protocols. *Learning Disabilities Quarterly, 16,* 19–32.

Office of the National Education Commission, Ministry of Education, Thailand. (2004). *Education in Thailand.* Bangkok, Thailand: Author.

Overtoom-Corsmit, R., Dekker, R., & Span, P. (1990). Information processing in intellectually highly gifted children by solving mathematical tasks. *Gifted Education International, 6,* 143–148.

Polya, G. (1957). *How to solve it: A new aspect of mathematical method.* Garden City, NY: Doubleday & Company, Inc.

Posamenteir, A. S., & Salkind, C. T. (1996). *Challenging problems in geometry.* New York: Dover Publications.

Posamenteir, A. S., & Schulz, W. (1996). *The art of problem solving: A resource for mathematics teacher.* Thousand Oaks, CA: Corwin Press, Inc.

Pressley, M., Borkowski, J., & Schneider, W. (1989). Good information processing: What it is and how education can promote it. *International Journal of Educational Research, 13,* 857–867.

Pretz, J. E., Naples, A. J., & Sternberg, R. J. (2003). Recognizing, defining, and representing problems. J. E. Davidson, & R. J. Sternberg (Eds.), *The psychology of problem solving,* (pp. 3–30). Cambridge, MA: Cambridge University Press.

Pugalee, D. K. (2001). Writing, mathematics, and metacognition: Looking for connections through students' work in mathematical problem solving. *School Science and Mathematics, 101,* 236–245.

Schoenfeld, A. H. (1985). *Mathematical problem solving.* Orlando, FL: Academic

Schoenfeld, A. H., Burkhardt, H., Daro, P., Ridgway, J.,Schwartz, J., & Wilcox, S. (1999). *High school assessment.* White Plains, NY: Dale Seymour Publications.

Sowell, E. J., Zeigler, A. J., Bergwell, L., & Cartwright, R. M. (1990). Identification and description of mathematically gifted students: A review of empirical research. *Gifted Child Quarterly, 34,* 147–154.

Sriraman, B. (2003). Mathematical giftedness, problem solving, and the ability to formulate generalizations. *The Journal of Secondary Gifted Education, 14,* 151–165.

Sternberg, R. J. (1995). *In search of human mind,* Orlando, FL: Harcourt Brace College Publishers.

Stevenson, H. W., & Stigler, J. W. (1992). *The learning gap: Why our schools are failing and what can we learn from Japanese and Chinese education.* New York: Summit.

Swanson, L. H. (1990). Influence of metacognitive knowledge and aptitude on problem solving. *Journal of Educational Psychology, 82,* 306–314.

Thipatdee, G. (1996). *The construction of an enrichment curriculum developing complex thinking ability of the upper secondary school students with high achievement.* Unpublished Doctoral Dissertation, Chulalongkorn University, Bangkok, Thailand.

Wieczerkowski, W., Cropley, A. J., & Prado, T. M. (2000). Nurturing talents/gifts in mathematics. In K. A. Heller, F. J. Monks, R. J. Sternberg, & R. F. Subotnik (Eds.), *International handbook of giftedness and talent education,* (pp. 413–425). Oxford, UK: Pergamon.

Yimer, A. (2004). *Metacognitive and cognitive functioning of college students during mathematical problem solving.* Unpublished Doctoral Dissertation, Illinois State University.

Yin, R. K. (1994). *Case study research design and methods,* Newbury Park, CA: Sage.

ACKNOWLEDGMENT

Reprint of Pativisan, S., and Neiss, M. (2007). Mathematical Problem solving processes of Thai gifted students. In B. Sriraman (Guest Editor), *Mediterranean Journal for Research in Mathematics Education*, 6(1 & 2), 47–68. Reprinted with permission from The Cyprus Mathematical Society. © 2007 Supattra Pativisan.

CHAPTER 9

KNOWLEDGE AS A MANIFESTATION OF TALENT

Creating Opportunities for the Gifted

Alexander Karp
Teachers College, Columbia University

ABSTRACT

This article discusses the link between mathematical giftedness and deep knowledge of mathematics. The author examines observations by teachers from schools for gifted students and the biographies of outstanding mathematicians. The link between talent and knowledge is analyzed both from a theoretical viewpoint and from a practical perspective, indicating what can be done to identify mathematically gifted students and to develop their talent. In carrying out this analysis, the author relies on the Russian experience of working with mathematically gifted students.

Creativity, Giftedness, and Talent Development in Mathematics, pages 209–224

INTRODUCTION

Talent and knowledge are often placed in opposition to one another. The Romantic tradition has left behind many images of "idle loafers" (Pushkin, 1954) whose natural gifts alone allow them to achieve what knowledge cannot. In fiction about schools, it is not uncommon to come across descriptions of situations in which it is precisely the failing student—who seemingly knows nothing—who turns out to be the most talented and creative in the class (e.g. see the old, but in many respects representative Russian example by Glushko, 1953). The existence of a tension between knowledge and creativity is noted in scientific literature as well. The present article, by contrast, will discuss evidence of broad and profound knowledge in many mathematically gifted students. The discussion will draw upon both the results of interviews conducted by us with teachers and biographical analysis of a number of outstanding mathematicians. The article will also analyze certain aspects of the Russian experience of working with the gifted and formulate certain propositions.

KNOWLEDGE, CREATIVITY, GIFTEDNESS: RESEARCHING AN UNEASY RELATIONSHIP

Weisberg (2000) notes:

> while it is universally acknowledged that one must have knowledge of a field if one hopes to produce something novel within it, it is also widely assumed that too much experience can leave one in ruts, so that one cannot go beyond stereotyped responding. The relationship between knowledge and creativity is assumed, therefore, to be shaped like an inverted U, with maximal creativity occurring with some middle range of knowledge. (p. 226)

One work that may be cited in support of such views is Simonton (1984), which investigated the relationship between individuals' achievements and their level of formal education. At the same time, other studies (Hayes, 1989; Weisberg, 2000) indicate that only sufficiently deep knowledge makes genuinely creative achievements possible. Moreover, Weisberg challenges certain earlier studies, noting in particular that greater formal education does not necessarily correspond to greater knowledge (this is especially true of situations that existed hundreds of years in the past—the ways of obtaining an education then were, and at least may have been, completely different from the ones common at present). Weisberg reaches the conclusion that the relationship between creativity and knowledge must be re-thought. Let us note that, in order to do so, we must define more precisely what we mean by both knowledge and creativity (e.g., see Sriraman, 2004).

Without going into further discussion of the relevant theories here, note that it is possible to study various aspects of this relationship. The foregoing discussion concerned for the most part *the influences* of knowledge on creativity: how increased knowledge affects creativity. Below, we will be mainly interested in knowledge as *a manifestation* of giftedness.

This topic, too, is the focus of theoretical controversies. Several approaches to defining the very notion of mathematical giftedness have been proposed (see Sriraman, 2005). It is clear that such a definition is a conventional one and depends upon whom we choose to include among the gifted. In general, it is natural to consider a person who has defended a doctoral dissertation in mathematics more gifted than someone who is unable to pass a state-mandated secondary school exam, despite, say, numerous lessons with a private tutor. At the same time, it may turn out to be impossible to consider the aforementioned doctoral degree recipient a highly gifted individual when measured against, say, Gauss. For the sake of convenience, Usiskin (2000) identified eight levels of giftedness (note that all such distinctions are by definition not precise: one could make do with seven levels, but one could also define nine, and the borders between the levels in each case are quite nominal). It is evident that the descriptive model will change in accordance with who is considered to be gifted. This will also influence the degree to which it will be desirable to link giftedness with creativity.

With a broad conception of giftedness—one which includes among the gifted all those who, for example, were able to obtain a master's degree in mathematics—it would be naive to consider creativity to be a necessary characteristic of all gifted individuals. But it is natural to do so if a more narrow conception of giftedness is being used. One of the most widely used definitions of giftedness is that of Renzulli, according to which "giftedness consists of an interaction among three basic clusters of human traits: above-average general abilities, high levels of task commitment, and high levels of creativity" (Ridge & Renzulli, 1981, p. 204). The discussion below will rely on a "narrow" conception of mathematical giftedness and consequently on Renzulli's model. In the specific instances of mathematical giftedness examined below, however, it will be useful to speak about specific mathematical abilities rather than merely general ones.

ON THE CHARACTERISTICS OF THE
MATHEMATICALLY GIFTED

The Russian psychologist Krutetskii (1976) carried out a fundamental analysis of mathematical abilities. His work included experimental investigations of problem solving by the gifted, longitudinal studies of different groups of students, and questionnaire-based research. This last aspect of his study included interviews with mathematics teachers aimed at determining "what teachers mean by an ability to learn mathematics, what criteria they use to judge ability, which people they regard as capable or incapable and why" (p. 82). Subsequently, written questionnaires were submitted to a different group of teachers. Based on the answers submitted by the mathematics teachers, a number of criteria and attributes of mathematical abilities was identified. Among them were:

1. Comparatively rapid assimilation of mathematical knowledge, skills, and techniques. Quick comprehension of the teacher's explanations.
2. The capacity for logical and independent reasoning.
3. Inventiveness and competence in finding solutions.
4. Rapid and lasting memorization of mathematical material.
5. A highly developed capacity for the generation, analysis, and synthesis of mathematical material.
6. Mental flexibility, and others.

In numerous subsequent articles, the named characteristics were given more precise definitions and their number increased (see Sriraman, 2005, for references). At least two of the characteristics identified by Krutetskii are directly related to students' knowledge. According to the teachers, gifted students have a natural predisposition to accumulate mathematical knowledge—they assimilate new material with ease, and they retain it with ease. The object of Krutetskii's study, however, is not students' command of special knowledge that goes beyond the framework of the school program.

A questionnaire survey of research mathematicians also conducted by Krutetskii revealed a tendency among them to put knowledge and the capacity for original thought in opposition to one another: "the difference between two types of mathematical minds is that some are quick to grasp and master alien ideas (they become learned persons, while other think more originally but more slowly" (p. 191).

INTERVIEWS WITH RUSSIAN TEACHERS: BACKGROUND AND METHODOLOGY

The study that will be discussed below possessed elements that resembled Krutetskii's study, both in terms of its aims and in terms of its methodology. In the present study, as in the older one, mathematics teachers were asked to tell about their most gifted students and to name their principal distinguishing characteristics (the interviews with the teachers also probed other issues, and these other findings are reported on in other articles).

Crucially, however, by contrast with Krutetskii's study, the participants of the interviews described here were not teachers in ordinary schools—whose encounters with highly gifted students are, in the end, rather rare—but teachers in specialized schools, also known as schools specializing in the study of mathematics.

Such schools have existed in Russia (the Soviet Union) since the early 1960s (Vogeli, 1997). Typically, in order to enroll in them, students must pass a series of exams. Their mathematics curriculum is usually considerably broader and deeper than that of ordinary schools (e.g. see Karp, 1992). Over the years, these schools have established a solid reputation for themselves, and it is fair to say that usually students who are interested in mathematics and have a talent for it enroll specifically in schools of this type. To illustrate their effectiveness, it may be noted that graduates from such schools make up over 90% of the faculty of the St. Petersburg University Mathematics Department (Donoghue et. al., 2000).

We conducted interviews with top teachers from such schools in Moscow and St. Petersburg (12 persons in all). The following criteria were used in selecting participants for the study: First, we took into account how many of each teacher's students had gone on to take part in or win high-level mathematics Olympiads; almost all of those interviewed had had participants and winners of International Olympiads among their students. Second, we considered how many of their former students had gone on to become prominent mathematicians (research mathematicians holding positions at top universities); practically all of those interviewed could name at least three former students who were now senior faculty in the mathematics department of a leading academic institution. Finally, the third principle relied on in selecting participants was their professional activity, as expressed in their number of professional publications; almost all of those interviewed had written no fewer than three such works. The vast majority of those interviewed displayed achievements in all three categories; however, in a small number of cases it was deemed acceptable as well to include those who possessed high achievements in only two of these categories. It should be noted that the number of mathematics teachers who have been working at such schools for a comparatively long time is in general low, and although it cannot be said that the present study included all of the top teachers, it is at least clear that a significant portion of them was indeed included, and most importantly, that all of those included are undoubtedly distinguished by striking achievements in their field. It is likewise clear that the teachers who were selected possessed unique experience in interacting with highly-gifted students. All of the interviews were audiotaped.

Some Results of the Interviews

In many respects, the interviews confirmed Krutetskii's findings. Very many teachers emphasize such characteristics of the highly gifted student as *success at problem solving and exceptional precision and depth of the student's understanding of the solution.* One teacher formulated this in the following way: "He solves any problem. He has to think for a minute and then he gives the solution. Naturally, this makes a big impression." Another teacher related the following incident:

> They were distinguished by the speed with which they reacted. How they grasped problems, their depth. They could take a problem apart completely. I remember I had a boy—no matter what I asked, he gave an instant solution. I picked and chose, made up some problem, and called him to the blackboard. He stands there for a minute—the blackboard is empty. Another minute—the blackboard is empty. Five minutes pass—the blackboard is still empty. And then, suddenly, he picks up the chalk, writes down literally two extremely short lines, and gives the answer. The person thinks, he literally immerses himself completely in his thoughts, and then—one, two, three, and done—the answer is ready.

Speed in problem solving, although it was remarked upon by many of the teachers interviewed, did not in and of itself constitute the key factor in mathematical giftedness, in their opinion: "slow people can also be very strong." Moreover, some of the respondents specifically distinguished between mathematical giftedness and success in mathematics Olympiads, emphasizing that not all winners of Olympiads could, in their view, be included among the most gifted students, and conversely, that some of the most gifted students did not attain the uppermost heights in the mathematics Olympiads (although they undoubtedly performed quite successfully). The crucial factor, in the opinion of these teachers, is *interest in problem solving.* "They are interested in solving the problem—not getting an A, not winning some kind of prize, but in solving the problem."

Another important characteristic displayed by highly gifted students is the *nonstandard* character of their solutions and their capacity for *independent thinking.* One of the teachers interviewed expressed this in the following way:

They virtually always come up with a different solution than the one I had in mind... Say, I'm thinking that a certain equation can be solved in such-and-such a way... So they say, "But that means that it's like this"—and give a geometric interpretation.

Crucially, along with the traits already mentioned, many teachers emphasize the *ability to assimilate a large quantity of mathematical material deeply and early on* as an important characteristic of the highly gifted student. In the words of one of the teachers:

> There was one indication for those who were truly "stars." The symptom was, by eighth or ninth grade, they knew an exceptionally large amount of things. In other words, an ability quickly to absorb completely unfamiliar material. And easily. And to retain it. In other words, they already knew a great deal— all kinds of abstract things. This wealth of knowledge was astonishing.

Another teacher provides a typical example of the way in which such knowledge is formed:

> I asked him what he was most interested in. He says: "Well, in the end, it has to be algebra." So I call up the head of the algebra department at the university. He says: "Yes, I have a seminar on Galois theory, but for graduate students." I say: "Sasha, I have a book on Galois theory, maybe you can look at it." So he read it and went to the seminar. After the seminar, he comes back and I say to him: "So, Sasha, how was it?" He says: I understood almost everything." I say: "What about the book?" He says: "Oh, with the book, I had no problem at all. I understood everything." That's how long it took—one week. After one week, the boy is ready to attend a seminar on Galois theory, to attend a seminar for graduate students in the mathematics department.

Another teacher told about a student of his whose abilities were first recognized when he was still attending an ordinary middle school, prior to enrolling at a school specializing in the study of mathematics. In addition to going to school, this student was also taking classes in mathematics by mail (at a so-called mathematical correspondence school). When this student sent in solutions to problems that he had solved, he also asked various, rather complex questions about integrals. It turned out that he was doing his brother's homework—and that his brother was a college student. Similar stories were told by other teachers. For example, another person interviewed told about a pupil of his who, from second grade on, started skipping school in order to stay at home and watch lectures on Calculus on television, which were then part of the educational television programming. As a result, by the time he enrolled in a school specializing in the study of mathematics (at age 14), he had a firm grasp of mathematics at approximately the Calculus III level.

CHILDHOOD MANIFESTATIONS OF VAST KNOWLEDGE
IN OUTSTANDING MATHEMATICIANS

The biographies of outstanding mathematicians contain vivid instances of vast mathematical knowledge in childhood. The self-described "ex-prodigy" Norbert Wiener (1964), who mastered the standard curriculum in higher mathematics at an extremely early age, can serve as a textbook example. The speed and depth with which Pascal mastered Euclid's geometry (Bell, 1937, p. 75) also became legendary. Hadamard received the highest score on the placement exams for both of the most prestigious French institutions of higher learning of his time—the École Polytechnique and the École Normale—which would have been impossible without extensive knowledge of mathematics. The same thing happened with Darboux, Picard, Borel (Maz'ya & Shaposhnikova, 1998). To his biography of Gauss, Bell (1937) gave the following subtitle: "By twelve he dreams revolutionary discoveries, by eighteen achieves them" (p. XII)—and this was really the case; Gauss had mastered everything necessary for this at an extraordinarily early age.

The following episode from Sofia Kovalevskaya's memoirs (1978) can shed light on the specific nature of the highly gifted child's perception of mathematics. Kovalevskaya tells how, already as an 11-year-old, she would study the walls of her room which, for economy's sake, purely by chance had been wallpapered with pages from Ostrogradsky's mathematics textbook. She continues:

> Many years later, when I was already fifteen I took my first lesson in Differential Calculus from the eminent Petersburg professor Alexander Nikolaevich Strannolyubsky. He was amazed at the speed with which I grasped and assimilated the concepts of limit and of derivatives, "exactly as if you know them in advance." And, as a matter of fact, at the moment when he was explaining these concepts I suddenly had a vivid memory of all this, written on the memorable sheets of Ostrogradsky; and the concept of limit appeared to me as an old friend. (p. 123)

At the same time, it is evident that not all great mathematicians displayed extensive knowledge at such an early age. Hilbert did not demonstrate any extraordinary knowledge and was certainly no kind of prodigy. Hilbert's biographer Reid (1970, p. 6) writes that "mathematics appealed to him because it was 'bequem'—easy, effortless. It required no memorization"—but he did not distinguish himself in his classes in any way (particularly when compared to Minkowski, who attended the same school at the same time). As for Newton, although he did display his gifts at an early age by today's standards, for a long time he did not distinguish himself in school in any way (other than being able to stand up for himself and to give his tormentors a thrashing, Bell, 1937, p. 92). Many biographers note that Einstein, too, did not distinguish himself in school—and there are numerous other examples of this sort.

Naturally, in analyzing biographies, particularly biographies of scientists who lived hundreds of years in the past, it is necessary to take into account the incompleteness of the surviving evidence. We know about Pascal's childhood largely thanks to his sister (Bell, 1937), and Kovalevskaya left an account of her childhood herself, but such sources are not always available, and moreover the fact that no one around a child took note of his or her knowledge does not mean that such knowledge did not exist. Nonetheless, sufficiently reliable evidence exists to confirm the fact that extensive mathematical knowledge in childhood is not a universal characteristic of all highly gifted mathematicians (for example, Hilbert himself described his own childhood).

Discussion: Some Theoretical Considerations

There can be no doubt that Hilbert was capable of assimilating a great amount of mathematical knowledge: his whole career as a scientist corroborates this fact. Of course, it is possible to suppose that he did not possess this talent as a child, and that it blossomed only at a later age. One can point to examples that confirm that various kinds of mathematical abilities do not necessarily initially display themselves at an early age. However, based on the fact that we know that mathematics came to him easily in childhood, it is possible to propose another explanation (which can also be supported by biographical evidence): Hilbert, like Newton, did not display an early interest in mathematics. Using Renzulli's terminology, one might say that they lacked "high levels of task commitment." In other words, it may be argued that they could have known much more than they actually did, but saw no need to concentrate on mathematics until reaching a certain age.

Such an interpretation would suggest that a high level of early knowledge—such as described above—may be seen as a complex characterization of mathematical talent, indicating the presence of at least two of the three attributes by which talent is defined: such children demonstrate not only the ability to master mathematical material rapidly and durably, which was described by Krutetskii, but in addition a level of concentration on and interest in mathematics that enables them to acquire knowledge which goes far beyond the bounds of the ordinary.

It would be desirable to study in greater depth the interaction between the three aforementioned groups of characteristics that define mathematical talent, and to describe each of them in a more precise fashion. It is clear, for example, that far from all "mathematical prodigies," who possess a deep level of knowledge at a very early age, go on to become great mathematicians. For example, although the aforementioned middle school student who helped his brother with his integrals, as well as the elementary school student who learned calculus by watching television, both received doctoral degrees in mathematics, neither of them became a major research mathematician (based on their number of publications in top journals, for example).

The teacher of the first of these students explained this by the peculiar nature of that student's task commitment, which was apparent already during his school years. Even as a student, he was inclined to solve only those problems which could be solved comparatively quickly. "If a problem couldn't be solved in half an hour, he would drop it," the teacher noted. It is true, of course, that thanks to his phenomenal abilities, the student could accomplish a great deal in the course of a half-hour; for example, this level of task commitment was sufficient for him to complete the course in Calculus while he was still a student in middle school.

The teacher of the other aforementioned student explained his comparative lack of success in mathematics as being principally due to his lack of creativity: "he always had to do everything by the rules—but in applying the rules, in combining the rules, he had no equal."

Thus, extraordinary early knowledge does not automatically imply brilliant success later on. Nonetheless, it may be viewed as a factor that significantly contributes to the probability of such successes. At this point in the discussion, however, it is necessary to pass from questions of pure research to issues involved in actual teaching practice.

IDENTIFYING MATHEMATICAL TALENT

It is relatively easy to compare different levels of mathematical abilities retroactively, for example by analyzing the biographies of different mathematicians. It is far more difficult to make predictions, anticipating in advance the development of a student's endowments and bringing them to light. Without going into a detailed analysis of the various ways in which talent may be identified, let us merely underscore their inevitable limitations. For example, problems that, at a specific stage of study and under specific circumstances, one might imagine to be accessible only to persons with exceptional abilities may turn out to be not so inaccessible after all in practice.

Thorndike (1921) claimed that "problems of a certain degree of complexity and abstractness they [certain students] simply cannot solve, just as they cannot jump over a fence five feet high or lift a weight of five hundred pounds." The problems he referred to, however, were often enough problems in elementary arithmetic and algebra, which could be solved by practically all students who had gone through a strategically planned course of instruction (which is what let Krutetskii, 1976, and other sharply to criticize Thorndike's work).

The influence of social factors is usually extremely high, and the aim of practical work must consist precisely in making this influence as beneficial as possible. The ideal strategy would be one that offers all students who possess striking talents such opportunities as would allow their talents to flourish. Kovalevskaya's talent would have likely developed even if the walls of her room had not been covered with pages from a mathematics textbook. But in the case of Gauss, for instance, it is impossible not to agree with Bell (1937) that "it was only by a series of happy accidents that Gauss was saved from becoming a gardener or a bricklayer" (p. 219). These happy accidents included the fact that his teachers supplied him with books, as much as it lay within their power to do so.

Ramanujan is considered a classic example of a "pure genius" without any education—he was "discovered" by Hardy on the basis of a letter that he, an unknown clerk, had written, which contained new results. However, as Usiskin (2000) correctly point out, the matter is not quite so simple. Ramanujan at one point had gotten hold of a two-volume book, "A Synopsis of Elementary Results in Pure and Applied Mathematics," which covered the subjects of the examinations one had to pass at that time to receive an honors mathematics diploma from Cambridge University. It was, in fact, this book that provided Ramanujan with the opportunity to study mathematics.

Identifying mathematical talent involves identifying not so much that which already exists, as that which might yet come into being—"the zone of the student's proximal development," to use Vygotsky's (1986) terminology. Consequently, collaboration with a more highly developed partner may turn out to be decisive event for the potentially mathematically gifted student. Such a partnership can take shape first and foremost in the creation and demonstration of opportunities for the gifted student. Below, we will analyze certain aspects of the Russian experience of working with gifted students from this point of view. In this context, it is more fruitful to make use of a broader definition of giftedness than the one employed earlier—and let us once more emphasize that it is virtually impossible to define a student's level of giftedness in advance with any degree of precision.

Note, too, that it appears more useful to speak specifically of possibilities, not of rigid requirements. The conception of "knowledge" that has been emphasized above, and will be emphasized below, connects knowledge with the awakening of independent interest. In those cases when a student who has been evaluated as highly gifted is simply compelled, in one way or another, to attend advanced courses, such independence—which is precisely the distinguishing characteristic of greatest importance—is naturally absent. Without going into a detailed discussion of the advantages and shortcomings of a more rigid system of requirements here, let us confine ourselves to agreeing with Usiskin (2000), who, in the article already cited, expresses the view that it is not at all clear whether Ramanujan would have been better off sitting in an ordinary classroom than pursuing his studies on his own.

OFFERING OPPORTUNITIES

In noting the role of mathematical teams and clubs—without which, according to him, it is practically impossible for students to develop—Usiskin (2000) writes that "these activities bring together interested students from different grades in a single school and they create a culture that makes it permissible for a student to express great interest in mathematics" (p. 155). The same role of fostering a culture—or, perhaps it would be better to say, of helping to assimilate students into a culture—in which it is "permissible" to be interested in mathematics, can by played at least in part by books as well. Students ought to be given the opportunity to become aware of the fact that a great deal lies beyond the boundaries of that which they are required to know in school. This stimulates the students' natural curiosity, helps to create a more correct attitude toward mathematical knowledge, illustrating its open and inexhaustible nature, and demonstrates that particular attention will be shown to those students who are especially interested in mathematics.

When discussing the grand traditions of the mathematical Olympiads and the mathematics circles (including the Russian variety), which have undoubtedly been extremely important in identifying and attracting the mathematically gifted (Kukushkin, 1996), one must not forget about the more flexible and individualized system—which has existed alongside of these two traditions—of involving students in mathematics through the spread of books. It is interesting to note, by the way, that during the initial stages of the Russian Olympiad movement, much less attention was paid to the results of the competitions; for example, winners were forbidden from taking part in any further Olympiads. But is was considered important that the winner should have some interactions with professional mathematicians and receive a modest collection of mathematical literature (Fomin, 1994).

This way of thinking suggests a mission for the book, and the first level in fulfilling such a mission consists in including supplementary materials and chapters—optional and more difficult than the rest—directly into the textbook. Such supplementary sections would contain not merely one or two problems, but large sets of assignments or theoretical sections related to the main text, but substantially enriching and expanding it. Thus, for example, in the Russian textbook *Matematika-7*, edited by Dorofeyev (1997), every chapter ends with a brief section, "For Those Who Are Interested." Shary-gin's (1999) textbook contains 49 sections, eight of which are accompanied by an asterisk to indicate that they are part of the additional material. Some textbooks, such as Werner, Ryzik, & Hodot (2001), present virtually all of the materials on several levels of exposition.

The second level consists in using handbooks with supplementary chapters along with textbooks, to be employed in special classes with advanced study of mathematics, as well as for independent reading (e.g. see the handbook by Atanasyan et. al., 1996). The topics covered in such books are close to those covered in the main textbooks, but they are laid out using more difficult and deeper methods or theoretical considerations (for example, the aforementioned book by Atanasyan et. al., 1996, offers many substantive problems in geometry, which can be solved using the vector method, and introduces the topic of inscribed and circumscribed circles and their corresponding properties in a considerably detailed fashion).

Finally, the third level consists in the reading of popular literature. In this respect, Russian schoolchildren have an enormous variety of both Russian books and translations to choose from, which are addressed to those who are interested in mathematics, from Rademacher and Toeplitz's classic text (1957) to the numerous pamphlets in the series "Popular Lectures in Mathematics" published by Nauka Press and partly translated into English.

At a certain level of mathematical development, gifted schoolchildren can begin reading more fundamental books as well. But it is at the crucial earlier stages in the maturation of their mathematical talent that they must be provided with materials which will enable their talent to evolve and express itself. Naturally, new technologies, and first and foremost the internet, also open up new opportunities for working with gifted students—they now have a wealth of new sources to draw upon, not just books, to acquire new knowledge.

THE PROBLEMS OF MATHEMATICS TEACHER EDUCATION: SOME PRACTICAL PROPOSALS

The discussion above addressed the significance of optional possibilities for the gifted student—opportunities that lie outside the walls of the classroom. Nonetheless, the teacher's role also appears important in this context, since teachers can help their students to discover new sources of knowledge for themselves (as was done by Gauss's teacher, who, in doing so, confessed that "he is beyond me," Bell, 1937, p. 222); but they can also hinder them in their pursuit of knowledge—for example, by exhibiting a lack of respect for supplementary reading.

Stanley (1987) once wrote about the importance of selecting teachers capable of teaching gifted students. Let us frame the issue more broadly and suggest that it is important to prepare teachers who might not necessarily be able to teach the gifted (for example, by presenting new sections in mathematics or by posing problems for them to investigate), but who have the capacity to support and to counsel them. Lee Shulman (1986) once proposed an important concept: "pedagogical content knowledge." By this he meant the knowledge of "the most useful [for purposes of teaching] forms of representation of ideas, the most powerful analogies, illustrations, examples, explanations and demonstrations." Pedagogical content knowledge also implies the knowledge of what topics children find interesting, or difficult, as well as the knowledge of typical student misconceptions. Familiarity with literature that may be of use to gifted students, which exists in every country and in practically every language, ought to be considered an essential component of such pedagogical content knowledge—as should familiarity, at however rudimentary a level, with the problematics of identifying and teaching the gifted.

The justified and necessary struggle against rote memorization and mindless cramming sometimes becomes transformed in the mind of the beginning teacher into a call for reduced attention to knowledge in general. This in turn inevitably leads to instruction without any substance, which can only frighten the gifted student (and any student) away from mathematics. To prevent such an outcome, prospective teachers should become acquainted with examples of the ways in which mathematically gifted students construct their knowledge and use it for their further creative activity (to the degree that this topic has been studied at the present time).

Note that the necessity of familiarizing the future teacher with "the whole range of abilities" (to quote the requirements for those wishing to obtain a New York State teaching license) is recognized virtually universally; in practice, however, such familiarization often comes down to acquaintance with only one side of the entire spectrum. It would be desirable to include special courses devoted to teaching the gifted in the mathematics teacher education program (Evered & Karp, 2000). However, even when no opportunity for organizing such special courses exists, general courses on the methods of teaching mathematics must address these topics.

CONCLUSIONS

At present, a great deal remains unclear about the way in which mathematical talent functions and develops. This makes it all the harder—and all the more important—not to overlook it in everyday practice. The beginning of this article quoted to words of Pushkin's Salieri about Mozart—"an idle loafer." Acquaintance with Mozart's biography reveals that this characterization in no way corresponds to the actual facts. Deep knowledge, as we have attempted to show, represents an important, complex characteristic of mathematical talent as well. It is essential to create conditions in which such knowledge can be acquired by students.

REFERENCES

Atanasian, L. S., Butusov, V. F., Kadomzev, S. B., Shestakov, S. A., & Iudina, I. I. (1996). *Geometria. Dopolnitel'nye glavy k shkol'nomu uchebniku 8 klassa.* [Geometry. Supplementary chapters to the school textbook for the 8th grade]. M.: Prosveschenie.

Bell, E. T. (1937). *Men of mathematics.* New York: Simon and Schuster.

Donoghue, E. F., Karp, A., & Vogeli, B. R. (2000). Russian schools for the mathematically and scientifically talented: Can the vision survive unchanged? *Roeper Review. A Journal of Gifted Education.* 22(2), 121–122.

Dorofeev, G. V. (Ed.)(1997). *Matematika. Arifmetika. Algebra. Analiz dannyh. 7 klass. Uchebnik.* [Mathematics. Arithmetic. Algebra. Data Analysis. 7th Grade Textbook]. M.: Drofa.

Evered, L., & A. Karp (2000). The preparation of teachers of the mathematically gifted: An international perspective. *NCSMST Journal,* 5(2), 6–8.

Fomin, D. (1994). *Sankt-Peterburgskie matematicheskie olimpiady.* [St. Petersburg mathematical Olympiads]. St. Petersburg: Politehnika.

Glushko, M. (1953). *Na vsiu ghizn'.* [All life long]. Simferopol': Krymizdat.

Hayes, J. R. (1989). Cognitive processes in creativity. In J. A.Glover, R. R. Ronning, & C. R. Reynolds (Eds.), *Handbook of creativity* (pp. 135–145). New York: Plenum

Karp, A. (1992). *Daiu uroki matematiki.* [Math tutor available]. M.: Prosveschenie.

Kovalevskaya, S. (1978). *A Russian childhood.* New York: Springer-Verlag.

Krutetskii, V. A. (1976). *The psychology of mathematical abilities in schoolchildren,* (J. Kilpatrick & I. Wirszup, Eds.; J. Teller, Trans.). Chicago: University of Chicago Press.

Kukushkin, B. (1996). The Olympiad movement in Russia. *International Journal of Educational Research,* 25(6), 552–562.

Maz'ya, V., & Shaposhnikova, T, (1998). *Jacques Hadamard, a universal mathematician.* AMS-LMS.

Pushkin, A. (1954). *Mozart i Salieri. Polnoe sobranie sochinenii.* v. 3. (Mozart and Salieri. Complete Works.) M.:Pravda.

Rademacher, H., & Toeplitz, O., (1957) *The enjoyment of mathematics: Selections from mathematics for the amateur.* Princeton, NJ: Princeton University Press,

Reid, C. (1970). *Hilbert.* New York: Springer-Verlag.

Renzulli, J. (1977). *The enrichment triad model: A guide for developing defensible programs for the gifted and talented.* Wethersfield, CT: Creative Learning Press.

Ridge, H. L., & Renzulli, J. (1981) Teaching Mathematics to the Talented and Gifted. In Vincents J. G. (Ed). *The mathematical education of exceptional children and youth* (pp. 91–266). Reston, VA: National Council of Teachers of Mathematics.

Sharygin, I. F. (1999).*Geometria. 10–11. Uchebnik.* (Geometry. 10–11. Textbook.) M.: Drofa.

Shulman, L. S. (1986). Those who understand: knowledge growth in teaching. *Educational Researcher, 15*(2), 4–14.

Simonton, D. K. (1984*). Genius, creativity and leadership.* Cambridge, MA: Harvard University Press.

Sriraman, B. (2004). The characteristics of mathematical creativity. *The Mathematics Educator, 14*(1), 19–34.

Sriraman, B. (2005). Are giftedness and creativity synonyms in mathematics? A theoretical analysis of constructs. *Journal of Secondary Gifted Education, 17*(1), 20–36.

Stanley, J. (1987). State residential high school for mathematically talented youth. *Phi Delta Kappan, 68*(10), 770–772.

Thorndike, E. L. (1921) *The new methods in arithmetic.* Chicago: Rand McNally.

Usiskin, Z. (2000). The development into the mathematically talented. *Journal of Secondary Gifted Education, 11*(3), 152–162.

Vogeli, B. R. (1997). *Special secondary schools for the mathematically and scientifically talented. An international panorama.* New York: Teachers College Columbia University.

Vygotsky, L. (1986). *Thought and language.* Cambridge, MA: MIT Press.

Weisberg, R. (2000). Creativity and knowledge: A challenge to theorics. In R. J.Sternberg (Ed.) *Handbook of creativity* (pp. 226–253). Cambridge University Press.

Werner, A., Ryzik, V., & Hodot, T. (2001). *Geometria-8.* (Geometry-8.) M.: Prosveschenie.

Wiener, N. (1964). *Ex-prodigy: My childhood and youth.* Cambridge: MIT Press.

ACKNOWLEDGMENT

Reprint of Karp, A. (2007). Knowledge as a manifestation of talent: Creating opportunities for the gifted. In B. Sriraman (Guest Editor), *Mediterranean Journal for Research in Mathematics Education, 6*(1 & 2), 77–90. Reprinted with permission from The Cyprus Mathematical Society. © 2007 Alexander Karp.

CHAPTER 10

AN ODE TO IMRE LAKATOS

Or Quasi-Thought Experiments to Bridge the Ideal and Actual Mathematics Classrooms

Bharath Sriraman
The University of Montana

ABSTRACT

This paper explores the wide range of mathematics content and processes that arise in the secondary classroom via the use of unusual "counting" problems. A universal pedagogical goal of mathematics teachers is to convey a sense of unity among seemingly diverse topics within mathematics. Such a goal can be accomplished if we could conduct classroom discourse that conveys the Lakatosian (thought-experimental) view of mathematics as that of continual conjecture-proof-refutation which involves rich mathematizing experiences. I present a pathway towards this pedagogical goal by presenting student insights into an unusual counting problem and by using these outcomes to construct "ideal" mathematical possibilities (content and process) for discourse. In particular, I re-construct the quasi-empirical approaches of six 14-year old students' attempts to solve this unusual counting problem and present the possibilities for mathematizing during classroom discourse

Creativity, Giftedness, and Talent Development in Mathematics, pages 225–249

225

in the imaginative spirit of Imre Lakatos. The pedagogical implications for the teaching and learning of mathematics in the secondary classroom and in mathematics teacher education are discussed.

INTRODUCTION

Imre Lakatos' (1976) classic book *Proofs and Refutations* contains an imaginary account of classroom discourse between the students and the teacher in the "ideal" classroom. This rich discourse occurs in the historical context of the problem of classifying regular polyhedra and constructing a proof for the relationship between the vertices, faces and edges of a regular polyhedra given by Leonhard Euler as $V + F - E = 2$. The various protagonists of the book and the teacher passionately argue over the validity of definitions, explore and conjecture plausible pathways to a proof, and refute the validity of definitions and steps in a proof by producing pathological (monstrous) counterexamples. The discourse one finds in this book occurs in Lakatos' rich imagination but begs the question as to whether such a discourse is replicable in the classroom. Yet, nearly 30 years have passed without any practical follow to Lakatos' vision for discourse in the ideal classroom. In this paper, the teacher attempts to re-construct the Lakatosian classroom by reflecting on pathways student discourse could take based on their responses to an unusual counting problem. The essence of the Lakatosian method lies in paying attention to the casting out of mathematical pathologies in the pursuit of truth. Typically one starts with a rule and clearly identifies the hypothesis. This is followed by an exploration of the possibility of its truth or falsity. The process of conjecture-proof -refutation results in the refinement of the hypothesis in the pursuit of truth in addition to the pursuit of all tangential hypotheses that arise during the course of discourse.

There are instances of mathematics educators creating a classroom environment that is conducive to such discourse. An outstanding example is the two-year study conducted by Harold Fawcett in the 1930s. Fawcett (1938) was successful in structuring a two-year teaching experiment with high school students that highlighted the role of argumentation in choosing definitions and axioms and illustrated the pedagogical value of working with a "limited tool kit." The students in Fawcett's (1938) study created suitable definitions, choose relevant axioms when necessary and created Euclidean geometry by using the available mathematics of Euclid's time period. As a result, in this two-year teaching experiment each student in the classroom essentially created their own version of Euclidean geometry by choosing and defending their definitions, axioms, theorems and proofs. The glimpses of the discourse one finds in Fawcett (1938) illustrates that the Lakatosian vision of the ideal classroom can in fact be approximated in reality. Fawcett's (1938) study is often quoted in math-education papers that expound the value of teaching Euclidean Geometry or Fawcett's (1938) instructional approach that allowed for Euclidean Geometry to be created as opposed to learned from a textbook. Other good recent examples of classroom pedagogy that emphasized the value of classical construction tools and the role of argumentation in geometry are recent classroom studies in the Netherlands. In these studies school children utilized instruments such as the conic section drawer (van Maanen, 1992), while others performed classic geometric constructions following instructions on ancient manuscripts, e.g., a pentagon from an ancient Persian manuscript (Hogendijk, 1996). My observation is that *the* crucial element in all the successful "projects" in the study of geometry reported was the classroom teacher, who had the pedagogical skills and a deep interest in historical content. Other examples are the use of historic clubs in which students act-out or role-play or study a historic text such as the *Elements* (Brodkey, 1996) and engage in argumentation whist establishing the truth of a given proposition (Fawcett, 1938).

The Lakatosian vision of classroom discourse and Fawcett's (1938) exemplary example (and others quoted above) suggests that "mathematization" is indeed possible in the secondary classroom. By "mathematizing" I mean the "act of putting a structure onto a structure" (Wheeler, 2001, p. 51). A specific process that mathematization encompasses in the act of generalization where one imposes/discovers a generalization underlying varying problem situations (Sriraman, 2004a, 2004b, Sriraman & Adrian, 2004a). Wheeler (2001) pointed out the difficulty of teaching teachers to analyze and understand students' ability to mathematize, and the urgency to engage their students in mathematizing real problems by facilitating discourse. By real problems, Wheeler (2001) did not mean problems situated in a real-world context that can often be contrived, but problems that lead to non-trivial and "tractable" mathematics.

THE USE OF ATYPICAL COUNTING PROBLEMS

Counting can be regarded as an activity that distinguishes us from other species. The history of mankind can be traced through the primitive acts of sharing equal quantities onto to the development of numerals that abstract the physical act of counting, onto to the development of sophisticated place value number systems that furthered our progress onto modernity. The history of mathematics is peppered with numerous famous counting problems. Galileo puzzled over the cardinalities of the set of integers and the set of even integers. This problem eventually led to the construction of transfinite arithmetic by Cantor as a reconstruction of ordinary arithmetic (Rotman, 1977). One can boldly generalize and say, "Everyone can count!" (to an extent), which begs the question of why counting problems occur less and less frequently as students' progress onto the secondary grades. Curricular documents from organizations like the Australian Education Council (1990) and the National Council of Teachers of Mathematics (2000) are rather vague about the place of "counting problems" in the secondary mathematics curriculum. By "counting problems" I mean situations where a certain phenomenon is observed among numbers (usually positive integers), which naturally lead one to examine the reasons why such a phenomenon occurs or problems that require a basic number theoretic insight such as divisibility of a number, properties of remainders etc. These types of problems are sometimes stressed under the discrete math curricular strand, which encompasses basic topics from combinatorics. However the point I will try to stress in this paper is that one does not need a text or a curricular requirement in order to make use of counting problems in the classroom. In particular my intent two-fold:

1. To demonstrate the wide range of mathematics that become accessible in the secondary classroom via use of counting problems.
2. To present the possibilities for mathematizing via student insights into the problem.

Pedagogical Hopes/Goals and Its Reflections in the History of Mathematics

One of our universal curricular goals is to convey a sense of unity among seemingly diverse topics within mathematics. In order to show a pathway into achieving these aforementioned goals I present the outcomes of the use of one such "novel" problem and the ensuing mathematical possibilities resulting from student insights into the problem. These students are prototypical of above average high school students (13–15 year range) at a rural American high school enrolled in Accelerated Algebra I, a course for motivated students. Over the course of the school year, as the teacher of this class, I assigned problems that students solved in their journals. The pedagogical hope was to mediate conditions that allowed students to investigate open-ended problems over an extended time period, to encourage students to write out their solutions in depth, and to get them to reflect on their solutions. Students worked on these problems in their journals for 7–10 days and were interviewed subsequently. Another pedagogical hope was that students would be inspired to adopt a quasi-empirical methodology when tackling problems they had never encountered before in their schooling experiences.

This hope was inspired partly by following the lead of how instruction is typically tailored in science. Science is characterized by physical principles discovered or inferred via systematic observation, hypothesis generation and testing through experiment. For instance, in a high school science lab experiment the validity of a scientific principle is tested by performing a structured experiment, recording observations followed by the application of the appropriate regression techniques (or other means of data analysis) on the data to test the validity of the principle. Some innovative teachers would set up a structured science experiment, in which students gathered data and then tried to infer the principle that worked. Can this scientific method be somehow adapted to the learning of mathematics? Can teachers facilitate the discovery of mathematical generalizations and underlying principles by using structured problem situations that result in the adoption of a quasi-empirical methodology characterized by construction of particular cases and observations and eventually result in new mathematics? Can students, when perplexed with a problem unsolvable with their extant mathematical toolkits or repertoire be "pushed" to create new mathematical tools required to solve a problem? As the old adage says "necessity is the mother of invention" and the history of mathematics is characterized by this necessity for building new tools to tackle troubling problems. Besides Cantor's creation of transfinite arithmetic mentioned earlier, there are other historical examples that illustrate my point. For instance, Goldbach's (1742) conjecture that all even integers ≥ 4 can be expressed as the sum of two primes as yet remains unanswered, but as a consequence has resulted in the search and the creation of new mathematical machinery, both computational and theoretical to tackle this problem.

Finally, I cannot help but point out the phenomenal growth of mathematics and the many beautiful theorems of the 20th century as a consequence of the willingness of most mathematicians to use the Axiom of choice.[1] Barry Lewis, the 2003–2003 president of the Mathematical Association in the U.K recently gave a broad perspective on the value of tool building in mathematics. "It was no accident that when at the beginning of the last century Einstein needed different tools to look afresh at the space-time continuum, those tools had already been fashioned . . . in an abstract system with no possible practical use. At least that is what Gauss thought for he didn't even bother to publish his work" (Lewis, 2003, p. 426).

In summary, can we use the inherent structure and beauty of a mathematics problem to inspire an exploration of mathematics through student insights into the problem and actually conduct classroom discourse in the spirit of Lakatos' (1976) exemplary and imaginative discourse in the "ideal" classroom? Let us see . . .

THE PROBLEM

Consider the following problem (Gardner, 1997): Choose a set S of ten positive integers smaller than 100. For example I choose the set S = {3, 9, 14, 21, 26, 35, 42, 59, 63, 76}. There are two completely different selections from S that have the same sum. For example, in my set S, I can first select 14, 63, and then select 35, 42, Notice that they both add up to 77. (14 + 63 = 77; 35+ 42 = 77) I could also first select 3, 9, 14 and then select 26. Notice that they both add up to 26. (3 + 9 + 14 = 26; and 26 = 26) No matter how you choose a set of ten positive integers smaller than 100, there will always be two completely different selections that have the same sum. Make up sets of your own and check for yourselves. Why does this happen? Prove that this will always happen.

STUDENT PATHWAYS, INSIGHTS AND TOOL BUILDING

The problem is clearly intriguing and conveys a sense of the mystery of the integers. I encouraged students to randomly pick integers between 1 and 100 and we constructed several ten-element sets. I now give an account of and re-construct the quasi-empirical approaches of six 13–15 year old students as they attempted to solve the aforementioned counting problem in their journal. Student journal writings and interview vignettes are used to re-create student pathways and insights into this problem.

Devising Techniques to Control Problem Variability

One of the students in the class (Matt) was particularly adept at discovering different selections of numbers that yielded invariant sums. However, he did not believe that this always occurred and embarked on constructing a counterexample based on controlling how the digits were chosen in the problem. Matt first tried to control the variability of the problem by holding the number in the tens place constant and then varying the numbers in the units place and conjectured that since one is being forced to repeat one or more of the digits from 0 through 9 in picking the ten elements this results in two sums being the same. For instance one can set the tens place as 8, and then start picking the digits for the units place, and we have ten choices for the digit in the units place (namely the digits 0–9) but then in the process one is forced to repeat the digit 8 (since the ten integers have to be different). Matt's original conjecture was actually based on the set {90, 91, 92, . . . , 99} where one can select distinct digits until one has to select 99 where the 9 repeats. This according to him caused different two selections in a ten-element set to yield the same sum. He then conjectured that this wasn't the case if one picked a set with less than ten numbers and produced the 5 element set {3, 7, 12, 78, 69, 84}, with the number 78 crossed out, and a list of the sums: 3 + 7 = 10; 3 + 12 = 15; 3 + 69 = 72; 3 + 84 = 87; 7 + 12 = 19; 7 + 69 = 76; 7 + 84 = 91; 12 + 69 = 81; 12 + 84 = 96; and 69 + 84 = 153. Matt had crossed out the 78 because he had accidentally repeated the digit 7 and his scheme was to construct a set in which all the digits were unique. He concluded that this set illustrated his point of different selections yielding the same sum only when one repeated the digits. Matt constructed a five-element maximal set in the digits from 0 to 9 occurred only once in order to illustrate his scheme but he did not consider sums, which combined three or more numbers in his set. If he had done so, he would have noticed that 3 + 12 + 69 = 84, an element in his set, thus disproving his claim (Sriraman, 2004a).

This leads one to wonder whether students had unconsciously understood the requirement of different selections yielding the same sum as strictly sums of two numbers. Consider the following solution of John.

Discovering the Counting Rule for Subsets of a Set

John's plan was to take ten numbers and see if this "worked," i.e., if two different two sums that equaled each other could always be found. He was going to try this numerous times to "see that the sum always equals another number." So, John's plan was to model the problem and verify if this was true, as in the first two problems. John started by picking the set {1, 15, 13, 4, 6, 99, 20, 75, 86, 51}, and found the sum 13 + 86 = 99, which is an element of the set. He then found that 75 + 6 + 4 + 1 = 86, which is another element in the set. I thought that John's second sum was really elaborate and made a note to ask him if he had developed a certain method to construct the sums. The next set picked by John was {2, 10, 18, 25, 52, 49, 91, 1, 86, 98}, and he found the sum 86 + 10 + 2 = 98, an element of the set. Clearly John understood "different selections" the way it was intended in the fact that he picked three and four digit sums.

The third set picked was {90, 91, 92, 93, 94, 95, 96, 97, 98, 99}, which I thought was a very interesting set to pick, because it gave the largest sum possible for sets containing ten integers between 1 and a 100. This fact will come into play when we discuss the solution later. The two sums found by John in this set were, 91 + 92 = 183 and 90 + 93 = 183. In his summary of the problem, John wrote that he had solved the problem exactly as he had planned. "I took sets of ten three times, and it worked every time." He then wrote that he was unsure about why this happened but his guess was that "there were barely enough numbers for it to work out."

John was asked to clarify his thought processes during the interview but had to cancel his first appointment because of a tennis tournament. When he came to the interview his journal had a "second look" to the problem. Apparently John had decided to look the problem over again. In his second look, John constructed the set {2, 5, 6, 8, 10, 12, 14, 16, 19, 20}. He found the sums $6 + 8 = 2 + 12 = 14$; $2 + 10 = 12$, an element in the set; $8 + 2 = 10$, an element in the set; and $6 + 2 = 8$, another element in the set. John then picked a second set {1, 3, 5, 7, 11,13, 15, 17, 19}, and wrote "$5 + 3 + 1 = 9$, and $9 + 7 + 1 = 17$, etc." I noticed that John had first picked integers at random between 1 and 20, and then picked a set of odd integers between 1 and 20. He was clearly beginning to choose the integers carefully at this stage, probably with the hope of finding something. In the margin of his journal he added the numbers $1 + 2 + 3 + 4 + 5 + 6 + 7 + 8 + 9 + 10 = 45$ and he drew the chart (Figure 10.1):

Based on his sum and the chart (Figure 10.1) John concluded,

> I am thinking that it always happens because there are numerous amount of ways to get the answer. I made a chart and just adding two numbers, there are 45 ways. If you add three numbers there is an even greater chance.

I was at first perplexed by John's solution, and tried to make sense of John's concluding argument. John was essentially saying that one could calculate a very large number of possible sums, which was a very good observation. The reader should note that these students had not encountered notions of enumerating subsets of a given set, which plays a role in the solution. However John was beginning to devise a systematic way of counting all such subsets. In his chart, John had calculated the total number of two number sums in a set containing ten elements, which is 45, or ten choose two in combinatorial terms. He then realized that there were even more possibilities for sums

1	2, 3, 4, 5, 6, 7, 8, 9, 10
2	3, 4, 5, 6, 7, 8, 9, 10
3	4, 5, 6, 7, 8, 9, 10
4	5, 6, 7, 8, 9, 10
5	6, 7, 8, 9, 10
6	7, 8, 9, 10
7	8, 9, 10
8	9, 10
9	10
10	

Figure 10.1 John's representation of the Number Sum Problem.

containing three numbers, and based on this observation concluded that this phenomenon occurred because "there were numerous amount of ways to get the answer." John had generalized that a large number of sums were possible, by counting all possible two number sums, and then by noting that an even greater number of three number sums were possible. This scheme highlights the possibility of students independently constructing the rule that the number of elements of a k-element set is 2^k.

John did not play with the given starting conditions of the problem. In other words he only tried 10 element sets, unlike Matt who decided to play with the constraints and see what happened. There were others who followed a different line of experimentation, which is now revealed in the following attempt of Jim.

Pondering Over the Mystery of the Number System

Jim understood the problem "to be saying that out of any 10 number set of positive integers less than 100, you can pick 2 numbers to add up to a number in the set. You could also take 3 numbers . . . there is also the possibility of 2 numbers adding to a number on both sides." His plan to start the problem was to "randomly make a 10 number set, and find at least one combo, then make another one to make sure it always works." Jim's plan basically involved verification of the already stated fact in the problem. He constructed two sets which were {1, 33, 44, 71, 52, 27, 13, 69, 88, 97} and {3, 14, 23, 39, 40, 55, 67, 71, 83, 99} and found the sums $49 + 52 = 13 + 88 = 101$, and $23 + 14 + 3 = 40$ respectively. There seemed to be no obvious system that Jim was using to find sums that were equal. Jim did not make up any other sets but concluded, "the answer deals with the base 10 system. There are really only 10 numbers to choose from, anything higher can be represented as $10 + x$, its not like there is a new symbol like to stand for something. For example $11 = 10 + 1$, and $52 = 10 + 10 + 10 + 10 + 10 + 2$. He then wrote that every number between 1 and a 100 would be a "combo" (addition) of 10 and another digit from zero through nine.

I did not understand Jim's "combo" argument. It seemed that the operation of addition on the base ten symbols somehow explained to Jim why the sums worked. I hoped that the interview would yield some insights into what Jim was trying to say. This particular vignette has been condensed for the reader. To the reader that wished to skip it, the essence of Jim's argument was as follows. "It worked because the only symbols that were available came from the set {1, 2, 3, 4, 5, 6, 7, 8, 9, 10}" and all numbers between 1 and a 100 were some additive "combination" of these symbols, and this somehow ensured that two selections added up to the same sum. To the reader that would like to see how Jim came to the above conclusion, the vignette follows:

Vignette 1

S: It took me a while to think abut the base 10 system thing. I had a couple of number lines, like ten number sets, and I always found numbers to add up to the same number. Then, I like, *on my paper route was thinking about it, and then it suddenly occurred to me that there were only ten numbers to choose from.* Everything else can be expressed as 10 plus something. Like 11 = 10 + 1, and 50 = 10 + 10 + 10 + 10 + 10, and it is not like there are any new symbols, *like 11 doesn't have a new symbol, it is just 10 and 1.*

I: Okay, so you are saying that the only numbers to choose from are?

S: 1, 2, 3, 4, . . . , 10, all the other numbers are a variation of that form.

I: Okay, but how does that make two sums to be the same?

S: Because these are in order, and they are smaller than 50, so it is easy to see them.

I: What if you were to pick numbers at random? Like the ones I showed you in class?

S: Then *a good idea would be to organize them, after you have randomly picked them, from least to greatest, and then it is easy to add up,* instead of adding by jumping around and stuff, so then you find as many possibilities as you can. *Like you take the first number and go along the line see if it will add up to any number in the set. Then you go to the next number, and then add up two.*

I: So how many days did you spend on this problem?

S: Like thinking it out, maybe 5 days, and then I wrote up task 1 two days ago, then I wrote everything up last night, and that took about me about 45 minutes.

I: So you thought about it for 5 days?

S: Yeah, I've been thinking about it for a long time (laughing).

[*The interview was resumed the following day as Jim had to leave for an after school activity*]

Next day

I: Okay, here we are again.

S: I somewhat wrote down why I think this happens. I think its' because of the set 1 through 10 and stuff, because the only numbers you can get are, you choose from that set.

I: How does picking ten numbers at random relate to your ten number set with 1 through 10?

S: Cause this, say 88 is just a combination of two of these numbers or something. Like any number you can have is one of these or a combination of these. So everything you make has to be a combination of these.

(We are looking at 49 + 52 = 13 + 88)

I: Okay say, I take 49 from your set, what do you mean by a combination?

S: Take 4 and 9 and put them together and you have 49.

I: And the 52?

S: That's 5 and 2.

I: Now how does that relate to 13 and 88?

S: Cause they come from the same set, its not like a different thing you pull the numbers from. It's from the same 1 through 10 set.

I: I'm still not sure about what you are choosing?

S: You pick like the digits, you can get a number for each digit out of this set.

I: But how does that make two sums equal?

S: Like all the numbers in here can add up to each other, like 1 + 1 = 2, 1 + 3 = 4, and you can on like that.

I: Okay. But how does that make 49 + 52 = 13 + 88? How do you go from there to there?

S: This is just a more advanced way or something. Instead of just one digit, it has two.

Jim was trying very hard to explain his solution. He was suggesting that the numbers in a ten number set would involve some combination of the digits from zero through nine, but it was still unclear to me how this would lead to two sums being equal. He attributed it to the "base 10 system and how things add up." The reader may recall that Matt presented a similar argument which involved the digits from zero through nine, but in his case he tried to verify his conjecture by trying sets with less than ten numbers and by controlling the choice of digits in these sets. In Jim's case, he was trying to gain an insight into the problem by reflecting over the sums in the set {1, 2, 3, 4, 5, 6, 7, 8, 9, 10} and observing that all sums could be expressed in the form $10a + b$, where a and $b \in \{0, 1, 2, 3, 4, 5, 6, 7, 8, 9\}$. This led him to conclude that two selections yielded the same sum because of how addition worked on the finite number symbols. Unlike John and Jim, Jamie decided to experiment with the starting conditions of the problem in order to gain an insight into why these mysterious sums occurred.

Playing With the Given Hypothesis

In her journal, Jamie began the problem by writing,

[T]he problem is asking you to first have a set of ten positive numbers all smaller than 100. The problem states that there will always be two selections from S that will have the same sum. It does not matter what the numbers are. What I need to figure out is why this happens and give examples why it does.

The reader will note that in restating the problem, Jamie had translated proving to "give examples why it does."

In order to start the problem, Jamie decided to pick out ten integers for her "different sets of numbers." Then she would find selections that gave the same sum. She wrote, "I may have to try different additions before I find sets that work, but there will always be two different selections that will have the same sum."

Jamie tried three different sets, and in each case found two selections that gave the same sum. Jamie's first set was {1, 2, 16, 19, 25, 40, 45, 72, 75, 89}, which gave $1 + 2 + 16 = 19$, an element in the set. She also found $72+16+1= 89$, an element in the set. Her second set was {3, 12, 15, 30, 35, 50, 63, 74, 87, 99}, which produced $3 + 12 = 15$, an element in the set, and $87 + 12 = 99$, an element in the set. Her final set was {89, 90, 91, 92, 93, 94, 95, 96, 97, 98}, which yielded $90 + 92 = 89 + 93 = 182$, and $93 + 95 = 94 + 96 = 188$.

At this point Jamie concluded,

I don't know how to prove this will always happen. There is probably some kind of theory why it does though. I don't know what this theory is or how to figure out what it is. I do think this always happens because it may have something to do that we are using only 10 integers.

The reader will note that her selection of the ten integers became less random in her third set, where she chose consecutive integers from 89 through 98. She had not devised a system for finding equal sums. The author hoped that the interview would reveal why she had not pursued the problem any further.

Vignette 2

    S: I tried out different sets and it worked every time. I really couldn't figure out why it worked, a theory or anything.

    I: So you could always find two selections?

    S: Yeah, I tried out different sets, and it always worked. I don't really know why it worked? I couldn't really prove a theory or anything (sighing).

    I: So, you tried different sets of ten numbers each.

S: (Silence) I tried to figure out why it works?
I: What did you think about?
S: I thought maybe it has to do with like, that it can't be negative numbers and it has to be between one and a hundred.

Jamie said that she would look at the problem again and then talk about it. Jamie's second attempt involved construction of numerous sets. This time Jamie had modified the constraints of the problem as follows (see Figure 10.2).

A. Several sets with ten negative integers between –1 and –100.
B. Several sets of varying size, containing positive and negative integers between 1 and 100 and –1 and –100.
C. Several sets with fifteen positive integers between 1 and 200.
D. Several sets of varying size, containing positive integers between 1 and 100.

Note: In each of the four categories, Jamie tried to construct an argument to necessitate the acceptance of the hypothesis in the given problem. The arguments constructed by Jamie in her journal for each category were:

A. {–1, –13, –65, –72, –73, –86, –89, –90, –96, –99}
   –89 – 1 = –90; –73 – 13 = –86.

If the rules were to use ten negative numbers less than 100 it would also work. But this would not solve the (given) problem because we have to use positive numbers.

B. {3, 7, 25, 31}; {–17, 27, 52}

Negative and positive numbers are used. These sets are less than 10 (elements). Nothing works here!

C. {3, 17, 24, 74, 84, 91, 93, 96, 108, 14, 121, 135, 145, 157, 163}
   93 + 3 = 96; 91 + 17 = 108; 121 + 24 = 145.

It could still work if some of the guidelines are further changed. Instead of numbers less than 100 it could be numbers less than 200 and fifteen numbers instead of 10.

D. {1, 3}

Cannot work with two numbers because there is no way to get two selections that work.

{1, 2, 3, 4, 5}

It could work here since 1 + 2 = 3 and 1 + 3 = 4. But what if we had {72, 93, 94, 95, 96} it does not work.

Figure 10.2 Jamie's experimentation with the hypothesis in the problem.

Based on her experimentation with the given hypothesis in categories A, B, C, and D (Figure 10.2) Jamie concluded. "In order to always get the solutions that work, there has to be 10 or more integers that are positive and I don't think it matters what the actual numbers are, like it does not have to be strictly under 100, it could be more." Jamie used examples/counterexamples in cases where the given hypothesis was tweaked in order to justify the correctness of the given hypothesis. This is certainly an interesting approach. It was also noteworthy that she attempted the problem again and tried to vary the given constraints in order to get some insight. The next solution illustrates the construction of one particular counter-example for the problem where the hypothesis was changed to sums of a 4-element set and shows similarities to Jamie's line of thought.

Constructing a Particular "Pathological" Case

Hanna began the problem by rewriting the example given by the researcher/teacher. She then posed to herself the question, " Why does this happen in every set of ten positive integers smaller than a hundred?" In order to answer this question, Hanna decided to "start by choosing a couple of sets of ten positive integers less than 100 and see if two different selections equal the same." After this, she would "look at her results and try to understand why?"

The first set that Hanna made was {19, 7, 29, 30, 3, 21, 15, 16, 17, 28}, and she found the sum 16 + 3 = 19, an element of the set. Her second set was {1, 2, 3, 4, 5, 6, 7, 8, 9, 10} and she found the sum 2 + 3 = 5, an element in the set. Her third set was {2, 3, 5, 8, 4, 6, 7, 1, 10, 11}, where she found the sum 5 + 4 = 8 + 1 = 9. The fourth set was {11, 3, 19, 7, 30, 2, 4, 6, 5, 9} and she found the sum 19 + 6 + 5 = 30, an element in the set. She tried four more sets, each with ten elements, and always found two selections that gave the same sum.

I could not detect any refinement in how Hanna chose the integers. The first six sets that Hanna constructed had integers that were less than 20. She had one set that had the integers from 1 to 10. She had also not devised a systematic way of calculating all the possible selections that would yield two equal sums. Her conclusion as to why this phenomenon was occurring based on her work on the eight sets was as follows:

> Out of all the sets of ten positive integers that I picked, they all had 2 selections that added up to equal the same number. I think this might happen because you aren't using any negative numbers and because you are not subtracting, so you won't have negative numbers. I think this might happen because you can pick ten numbers under 100, so that gives you a wide variety to pick from. Also ten numbers gives you quite a few numbers to play with and

add to give you equal sums… *But if you could only have a set of 4 or 5 numbers under a 100, you wouldn't always get 2 equal sums because you have less numbers to deal with.* (Journal entry)

Hanna then constructed a four element set {19, 2, 48, 1} and calculated all the possible sums in this set besides the trivial sums of 19, 2, 48, and 1. The remaining sums were: 19 + 2 = 31; 19 + 48 = 67; 19 + 1 = 20; 2 + 48 = 50; 2 + 1 = 3; 48 + 1 = 49; 19 + 2 + 48 = 69; 19 + 2 + 1 = 22; 19 + 48 + 1 = 68; 2 + 48 + 1 = 51; 19 + 2 + 48 + 1 = 70 and concluded that "nothing in that set equals the same because there isn't enough numbers that you could put together to equal a certain sum that would equal other selections sum."

Her argument that there was a great deal of variety (in terms of possible selections) in a ten number set was plausible and similar to that of John's. She had conjectured that a smaller set would not have "variety" and hence two selections would not give the same sum in such a set. She supported her conjecture by constructing a four-element set in which two selections did not yield the same sum. This counterexample would fall into category 2 of Jamie's experimentation.

Having seen four approaches to the given problem, I finally present one very unusual approach devised by Amy.

Constructing an Unusual Set

Amy was intrigued by this problem and constructed an unusual set in the process of experimenting with various number sums. After constructing numerous ten-element sets, Amy was certain of the phenomenon of different selections yielding the same sum. She wrote in her journal that repeated verification of this phenomenon simply validated that it always worked and in order to find out why different selections yielded the same sum, a completely new approach was required in which each element was purposefully chosen. She constructed the set {1, 2, 4, 8, 16, 32, 64,…} and wrote

OK, I think maybe I got somewhere! OK, to choose this set I tried to make the most variety possible. Let me try to explain…I started out with 1, then I chose 2. Now, I didn't want to get a solution, so my next number obviously wasn't going to be 3, so I put 4 instead, because 1 + 2 = 3 and I would have had a solution already. So, I just continued working this way. The next bigger number would be 7, because 1 + 2 + 4 = 7, so I didn't choose 7, instead I chose the next number 8. Now, 1 + 2 + 4 + 8 = 15, so I wouldn't want to pick 15, because that would be a solution, so I chose 16. When I had these five numbers {1, 2, 4, 8, 16}…I discovered a pattern. I just had to keep doubling the last number to get the next number… (and got) the set {1, 2, 4, 8, 16, 32, 64,…}. (Sriraman, 2003)

Once this set was constructed, Amy wrote that she couldn't continue with her scheme since the problem required that the elements in the set be between 1 and 100. She conjectured that her 7-element set was a "maximal" set, which did not yield selections of numbers that added to the same sum (Sriraman, 2004b).

THE LAKATOSIAN POSSIBILITIES OF DISCOURSE BY MATHEMATIZING STUDENT STRATEGIES/SOLUTIONS

Having presented six student pathways into this problem, we are now in a position to reflect on the nature of the outcomes and appreciate the mathematical possibilities (content and process) in the classroom. In this section I first present plausible classroom scenarios based on the actual student insights discussed in the previous section. In doing so I am putting myself in the shoes of the reflective teacher that analyzes student insights to recognize the rich mathematizing experiences in these student insights with the goal of facilitating classroom discourse that lead to non-trivial mathematics. One can indeed imagine the Lakatosian setting where these six students are present in the "ideal" classroom, engaged in classroom discourse based on their attempts of the given problem, where the teacher merely facilitates the mathematizing experiences that lead to non-trivial mathematics.

Figure 10.3 shows plausible classroom discourse based on student strategies and insights. Figure 10.4 construct possible pathways into diverse content based by mathematizing student strategies. This figure also contains elements of the Lakatosian process of conjecture-proof-refutation to lead to rich mathematics. In the next section both figures are discussed in depth in addition to a discussion of the actual mathematics arising from mathematizing student insights/strategies to fruition, i.e., discovery of general structures and deeper results.

Discussion of Plausible Discourse Pathways

The discourse begins in Figure 10.3 with the teacher asking the six students whether they were able to find a reasonable explanation for the mysterious phenomenon of different selections yielding the same sum in the given problem. Matt, who was especially an enthusiastic student in this course, does not agree this is possible based on his strategy of picking elements so that no digit repeats. Amy, Jamie and John who have verified this phenomenon for numerous 10-element sets point out that these sums occurred in all the 10-element sets they have constructed. Jamie thinks there is some theory that explains why this happens. John suggests that these

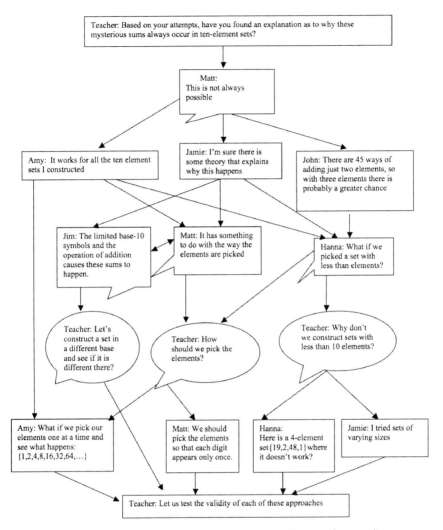

Figure 10.3 Plausible discourse based on student solutions and strategies.

sums occur because of the number of ways in which one can calculate such sums in any 10-element set, and says that there are 45 ways of combining two elements at a time. At this juncture there are several pathways that the discourse can take. Conceivably, Matt points out that these sums occur because of a particular method of choosing the ten elements. Hanna suggests that such sums would not occur if one picked a set with less than ten elements. Jim builds off Jamie and Matt's remarks and believes the sums to be a function of the finite way in which one can choose the ten elements using the digits of the base-10 numeral system. At this juncture the teacher can

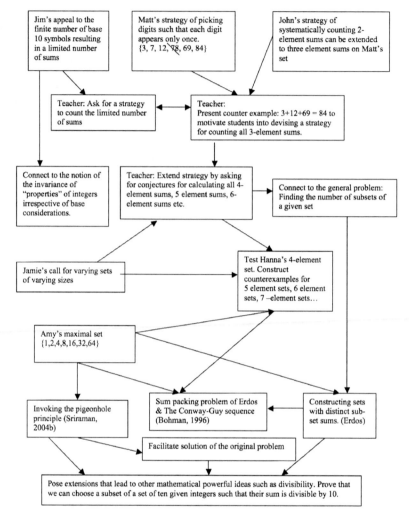

Figure 10.4 Pathways into mathematizing experiences via conjecture-proof-refutation in discourse.

pose several questions that will facilitate/encourage the students to share their examples with the class. For instance asking "How should we pick the elements" would allow Amy and Matt to share their insights of constructing a 7-element maximal set where no sums occur, and constructing a set in which no digit is allowed to repeat respectively. The teacher could also ask for examples of sets with less than ten elements, so that Hanna and Jamie can share their work with the others.

In order to further Jim's suggestion about the mysterious working of the base-10 system, the teacher can ask for sets of ten elements in a different base. The important thing at this stage would be to test the validity of all the conjectures/explanations by constructing the appropriate refutations and proofs in order to mathematize (and cash out!) the rich conjectures posed by the students. The teacher is also in a position to point out to the students their tendency of playing with the hypothesis is a natural practice among mathematicians in order to gain an insight into a given problem. The reader should note that although the discourse is constructed in the Lakatosian imaginative spirit, it is certainly very plausible since it is based on actual (factual!) responses of students in their journals and interview transcripts.

Discussion of Mathematizing Experiences Leading to Non-Trivial Mathematics

Figure 10.4 gives the rich possibilities of mathematizing student insights with the teacher facilitating (or orchestrating) this process. I also discuss the remarkable mathematics that arises from student insights into the problem. Matt's strategy of constructing a set in which each digit only occurs once {3, 7, 12, 78, 69, 84}, in order to refute the suggestion that such sums always occur can be re-refuted by the teacher by presenting the counter example: $3 + 12 + 69 = 84$ to motivate students into devising a strategy for counting all 3-element sums. This would allow John to share his strategy of systematically generating 2-element sums. Jim's appeal to the finite number of base 10 symbols resulting in a limited number of sums can also be now tested by asking students to estimate the number of sums possible in a ten-element set.

Students can be asked to conjecture a strategy for 3-element sums based on John's work and this can be extended indefinitely and nicely connected to the general problem of counting the number of subsets of any given set, thus illuminating the general structure of subsets of a set. The ambitious teacher can extend this further by asking students to construct sets with distinct subset sums, a problem posed by Paul Erdös in 1931, and pose the sum-packing problem of Erdös. The sum-packing problem of Erdös is as follows:

A set S of positive integers has distinct subset sums if the set $\{\Sigma_{x \in X} x: X \subset S\}$ has $2^{|S|}$ distinct elements. Let $f(n) = \min \{\max S: |S| = n$ and S has distinct subset sums$\}$. How small is $f(n)$?

In 1931, Erdös conjectured that $f(n) \geq c\, 2^n$ for some constant c, and in 1955 Erdös and Moser proved that $f(n) \geq 2^n/(10\sqrt{n})$ and this remains as the best estimate for a lower bound (Bohman, 1996). Naturally we ask ourselves what is the upper bound? If we take the set S to be the first n powers of 2 as Amy did, we can easily see that $f(n) \leq 2^{n-1}$. Amy's unusual insight into the problem and her "maximal" set construction actually yield the first three numbers of the Conway-Guy sequence constructed by Conway and Guy in 1967 (Guy, 1982) arising from this non-trivial sum packing problem. The problem (Gardner, 1997) given to the students can be generalized to get the Conway-Guy numbers as follows. As defined above, we let $f(n)$ be the smallest positive integer such that there exists n positive integers $\leq f(n)$ for which all subset sums are distinct. The first three values of f are $f(1) = 1$, $f(2) = 2$, $f(3) = 4$. Surprisingly enough $f(4)$ is not 8 but 7. The bound of 8 is suggested by the fact that the subset sums of the first four powers of two –1, 2, 4, and 8 are clearly distinct. This was Amy's insight when she tried to construct a set with distinct sums (Sriraman, 2004b). Amy's idea might lead the reader (the "ideal" class) to believe that if four numbers ≤ 7 are chosen, then two different selections from the set must have the same sum. This is shown false by considering the set $S =\{3, 5, 6, 7\}$ (Check it out!). Perhaps more surprising is the fact that $f(6) = 24$, not 32, which the reader might suspect by binary representation considerations. The corresponding set is $S = \{11, 17, 20, 22, 23, 24\}$. The first eight values of the Conway-Guy sequence $f(n)$ are 1, 2, 4, 7, 13, 24, 44, 84 . The upper bound found by Conway and Guy on S was 2^{n-2} provided n is large enough. Such a pathway would mathematize and illuminate the grand structure beneath the original problem.

A different pathway to the sum-packing problem and the Conway Guy numbers is via the strategy for systematically generating subsets, and hence all possible sums can be used to test Hanna's four-element set. A natural question is whether we can construct five-element sets, six-element sets, seven element sets and so on, with integers chosen from 1 to 100, such that different selections do not yield the same sum. This question would allow the teacher to take Amy and Jamie's ideas and mathematizate it to fruition by leading into the sum-packing problem. Finally, as mentioned earlier, Amy's seven element maximal set again naturally leads to the sum-packing problem of Erdos as well as the Conway-Guy numbers process with appropriate facilitation.

Another possibility for mathematization after students have constructed ways to generate subsets of sets and the number of such subsets is to have Amy invoke the pigeonhole principle (Sriraman, 2004b) in order to solve the original problem. The pigeonhole principle can then be used to facilitate the solution of the original problem by observing that the smallest possible sum is 1 and the largest possible sum is $90 + 91 + \ldots 99 = 945$, but the number of subsets of a ten element set is much larger than this, namely $2^{10} = 1024$, thereby forcing numerous selections of subsets to yield the same sum. This would merge the diverse ideas of the students from examining the set {90, 91,...99} (Matt) to efficiently calculating sums (John) to the hunch of the finite number of sums because of the limited choice of digits (Jim) to conforming the validity of the hypothesis for Hanna and Jamie. In fact now, Hanna, Jamie and the others could test the validity of their conjectures for the sets with less than 10 elements. This would confirm Jamie's hunch that a 15 element set with integers chosen from 1–200 would certainly yield numerous selections that add up to the same sum $[2^{15} > 190 + 191 + \ldots 199]$. There are myriad possible extensions that mathematize other powerful ideas such as divisibility and the wide-ranging applicability of the pigeonhole principle. For instance: Prove that we can choose a subset of a set of ten given integers such that their sum is divisible by 10 (Fomin, Genkin, & Itenberg, 1996).

CONCLUSIONS AND IMPLICATIONS

The preceding discussion highlights the plethora of mathematics that becomes accessible to secondary students if the teacher makes use of student insights on atypical problems to lead students to discover mathematical structures. Although the six students only had the mathematical sophistication of beginning Algebra students and hence operating with a limited toolkit, their attempts revealed the natural tendency to create new mathematical tools to tackle this problem. In a sense their mathematical behavior was analogous to that of the rich tool building process characterizing the history of mathematics when mathematicians were confronted with perplexing problems such as Fermat's Last Theorem and The Four-Color Problem. Thus, creating mathematical experiences that necessitate the creating of new tools is a very useful pedagogical technique.

The value of choosing novel (atypical) problems that capture students' interests and seem accessible to students needs to be emphasized to practitioners. Numerous studies seem to indicate that combinatorial (e.g., English, 1998, 1999; Hung, 1998; Maher & Kiczek, 2000; Maher & Martino, 1996a, 1996b, 1997; Maher & Speiser, 1997; Sriraman, 2004a, 2004b, 2004c) and number theoretic problems (Sriraman, 2003b, 2004d ; Sriraman & Strzelecki, 2004a) are especially useful in serving the purpose of being accessible as well as lead into investigations of the underlying structure. The accessibility of simply stated (but mathematically complex) problems help teachers to foster independent thinking in the classroom (Sriraman & English, 2004a).

Although the current trend in mathematics education is to emphasize model-eliciting problems/activities (Lesh & Doerr, 2003) situated in a real-world context as a means to catalyze mathematizing in the classroom, novel pure math problems still have an important place in the curriculum.

Encouraging students to tackle atypical counting problems to make deep connections with topics in Number Theory, Combinatorics and Analysis complements the Applied mathematics and Statistics that students learn through the modeling approach that is presently gathering momentum. The underlying hope is that mathematics educators never forget the aesthetic beauty inherent in pure math activities and convey to our students that such activities have sustained the imagination of mathematicians and contributed to its growth from the very onset of its history. (Sriraman & Strzelecki, 2004a)

The value of allowing students extended time periods to work on problems and encourage them to engage in reflective journal writing serves multiple purposes. It not only creates a non-threatening medium via which students are willing to try multiple strategies, but it also helps the teacher to plan lessons that involve mathematical discourse with the goal of facilitating the creation of new mathematics as well as the discovery of structure. Student journal writings are also an invaluable asset to initiate and orchestrate classroom discourse with the aim of mathematizing in mind. Even incorrect student attempts or counter-examples serve the pedagogical purpose of allowing the teacher or other students to construct appropriate refutations that will allow students to make the necessary "tweak" to move the mathematics forward. The preceding student attempts on the given problem seem to indicate that students have a pre-disposition to experiment with the hypotheses when the problem is too difficult in the stated form. After all, new mathematics is created by this continual "tweaking" process in which preliminary hypothesis undergo refinement until a theorem emerges. The Lakatosian methodology of conjecture-proof-refutation conveys a vibrant and alive picture of mathematics. It would be worthwhile to expose prospective teachers by modeling this process in mathematics and mathematics education courses. The preceding hypothetical classroom discourse constructed with actual student insights reveals the rich mathematics that becomes accessible via the Lakatosian methodology and careful facilitation of discourse. This also implies the necessity to impress upon prospective teachers the value of an in-depth knowledge of graduate level mathematics.

Doerr and Lesh (2003) recently called for mathematics educators to recognize the applicability and the analogs of Dienes (1960, 1961) principles to teacher education. For instance Doerr and Lesh's (2003) "multilevel principle" is the instructional analog of Dienes (1960, 1961) "dynamic principle" which Doerr and Lesh emphasize by saying "Teachers most often need to simultaneously address content, pedagogical strategies, and psychological aspects of a teaching and learning situations" (p. 133). This implies that when we give students an open-ended problem, it becomes our responsibility to address as many aspects of the problem (as possible) and ideally mathematize it to fruition by leading into the discovery of structure as demonstrated earlier. In this endeavor the Lakatosian methodology of conjecture-proof-refutation serves as a valuable tool to mathematize unusual problems and bridge the "ideal" Lakatosian classroom with the actual mathematics classroom!

REFERENCES

Australian Education Council (1990). *A national statement on mathematics for Australian schools.* Melbourne, VC: Australian Educational Council.

Bohman, T. (1996). A sum packing problem of Erdös and the Conway-Guy sequence. *Proceedings of the American Mathematical Society, 124*(12), 3627–3636.

Brodkey, J. J. (1996). Starting a Euclid club. *Mathematics Teacher, 89*(5), 386–388.

Dienes, Z. P. (1960). *Building up mathematics.* London: Hutchinson Education.

Dienes, Z. P. (1961). *An experimental study of mathematics learning.* New York: Hutchinson.

Doerr, H., & Lesh, R. (2003). A modeling perspective on teacher development. In R. Lesh & H. Doerr (Eds.), *Beyond constructivism* (pp. 125–140). Mahwah, NJ: Erlbaum.

English, L. D. (1998). Children's problem posing within formal and informal contexts. *Journal for Research in Mathematics Education, 29*(1), 83–106.

English, L. D. (1999). Assessing for structural understanding in children's combinatorial problem solving. *Focus on Learning Problems in Mathematics, 21*(4), 63–82.

Fawcett, H. P. (1938). *The nature of proof. Thirteenth yearbook of the NCTM.* New York: Bureau of Publications, Teachers College, Columbia University.

Fomin, D., Genkin, S., & Itenberg, I. (1996). *Mathematical Circles (Russian Experience).* American Mathematical Society.

Gardner, M. (1997). *The last recreations.* New York: Springer-Verlag.

Goldbach, C. (1742) Letter to L. Euler, June 7, 1742. (Accessed on January 11, 2004) http://www.mathstat.dal.ca/~joerg/pic/g-letter.jpg

Guy, R. K. (1982). Sets of integers whose subsets have distinct sums, *Theory and Practice of Combinatorics, Annals of Discrete Math, 12,* 141–154. North-Holland, Amsterdam.

Hogendijk, J. P. (1996). Een workshop over Iraanse mozaïken. *Nieuwe Wiskrant, 16*(2), 38–42.

Hung, D. (1998) Meanings, contexts and mathematical thinking: the meaning-context model. *Journal of Mathematical Behavior, 16*(3), 311–344.

Lakatos, I. (1976). *Proofs and refutations.* Cambridge, UK: Cambridge University Press.

Lesh, R., & Doerr, H. (2003). Foundations of a models and modeling perspective on mathematics teaching, learning and problem solving. In R. Lesh and H. Doerr (Eds.), *Beyond Constructivism* (pp. 3–34). Mahwah, NJ: Erlbaum.

Lewis, B. (2003). Taking perspective. *The Mathematical Gazette, 87*(510), 418–431.

Maher, C. A., & Kiczek, R. D. (2000). Long term building of mathematical ideas related to proof making. *Contributions to Paolo Boero, G. Harel, C. Maher, M. Miyasaki. (organisers) Proof and Proving in Mathematics Education.* ICME9 -TSG 12. Tokyo/Makuhari, Japan.

Maher, C. A., & Martino A. M. (1996a) The Development of the idea of mathematical proof: A 5-year case study. *Journal for Research in Mathematics Education, 27*(2), 194–214.

Maher, C. A., & Martino, A. M. (1996b) Young children invent methods of proof: The "Gang of Four." In P. Nesher, L. P. Steffe, P. Cobb, B. Greer & J. Goldin (Eds.) *Theories of mathematical learning* (pp. 431–447). Mahwah, NJ: Erlbaum.

Maher, C. A., & Martino, A. M. (1997) Conditions for conceptual change: From pattern recognition to theory posing. In H. Mansfield (Ed.), *Young children and mathematics: Concepts and their representations.* Durham, NH: Australian Association of Mathematics Teachers.

Maher, C. A., & Speiser, B. (1997). How far can you go with block towers? Stephanie's intellectual development. *Journal of Mathematical Behavior 16*(2), 125–132.

National Council of Teachers of Mathematics. (2000). *Principles and standards for school mathematics*: Reston, VA: Author.

Rotman, B.(1977). *Jean Piaget: Psychologist of the real.* Cornell University Press.

Sriraman, B (2003a). Mathematical giftedness, problem solving, and the ability to formulate generalizations. *The Journal of Secondary Gifted Education, 14*(3), 151–165.

Sriraman, B (2003b) Can mathematical discovery fill the existential void? The use of conjecture, proof and refutation in a high school classroom (feature article). *Mathematics in School, 32*(2), 2–6.

Sriraman, B. (2004a). Discovering a mathematical principle: The case of Matt. *Mathematics in School, 33*(2), 25–31.

Sriraman, B. (2004b). Reflective abstraction, uniframes and the formulation of generalizations. *Journal of Mathematical Behavior, 23*(2).

Sriraman, B. (2004c). Discovering Steiner Triple Systems through problem solving. *The Mathematics Teacher, 97*(5) 320–326.

Sriraman, B. (2004d). Re-creating the Renaissance. In M. Anaya, & C. Michelsen (Eds.), *Proceedings of the topic study group 21: Relations between mathematics and others subjects of art and science: The 10th International Congress of Mathematics Education* (pp. 14–19). Copenhagen, Denmark.

Sriraman, B., & Adrian, H. (2004a) The pedagogical value and the interdisciplinary nature of inductive processes in forming generalizations. *Interchange: A Quarterly Review of Education, 35*(4), 407–422.

Sriraman, B., & English, L. (2004a). Combinatorial mathematics: Research into practice. Connecting research into teaching. *The Mathematics Teacher, 98*(3), 182–191.

Sriraman, B., & Strzelecki, P. (2004a). Playing with powers. *The International Journal for Technology in Mathematics Education, 11*(1), 29–34.

Van Maanen, J. (1992). Teaching geometry to 11-year-old "medieval lawyers." *The Mathematical Gazette, 76*(475), 37–45.

Wheeler, D. (2001). Mathematisation as a pedagogical tool. *For the Learning of Mathematics, 21*(2), 50–53.

ACKNOWLEDGMENT

Reprint of Sriraman, B. (2006). An ode to Imre Lakatos: Bridging the ideal and actual mathematics classrooms. *Interchange: A Quarterly Review of Education.* (Special Issue of the International History, Philosophy and Science Teaching Group), *37*(1 & 2), 155–180. Reprinted with permission from Interchange & Bharath Sriraman, copyright holder of article. © 2006 Bharath Sriraman

NOTE

1. Suppose C is a collection of nonempty sets. Then one can *choose* a member from each set in that collection. Therefore, there exists a function f defined on C such that, for each set S in the collection, $f(S)$ is a member of S.

CHAPTER 11

THE MATHEMATICALLY GIFTED KOREAN ELEMENTARY STUDENTS' REVISITING OF EULER'S POLYHEDRON THEOREM

Jaehoon Yim, Sanghun Song, and Jiwon Kim
Gyeongin National University of Education, South Korea

ABSTRACT

This paper explores how the constructions of mathematically gifted fifth and sixth grade students using Euler's polyhedron theorem compare to those of mathematicians as discussed by Lakatos (1976). Eleven mathematically gifted elementary school students were asked to justify the theorem, find counterexamples, and resolve conflicts between the theorem and counterexamples. The students provided two types of justification of the theorem. The solid figures suggested as counterexamples were categorized as (a) solids with curved surfaces, (b) solids made of multiple polyhedra sharing points, lines, or faces, (c) polyhedra with holes, and (d) polyhedra containing polyhedra. In addition to using the monster-barring method, the students suggested two new types of conjectures to resolve the conflicts between counterexamples and the

Creativity, Giftedness, and Talent Development in Mathematics, pages 251–269
Copyright © 2008 by Information Age Publishing
All rights of reproduction in any form reserved.

251

theorem, the exception-baring method and the monster-adjustment method. The students' constructions resembled those presented by mathematicians as discussed by Lakatos.

INTRODUCTION

One perspective on mathematics education states that it is important to analyze and reconstruct the historical development process of mathematical knowledge for improving mathematics teaching and learning. A number of scholars including Clairaut (1741, 1746), Branford (1908), Klein (1948), Toeplitz (1963), Lakatos (1976), Freudenthal (1983, 1991), and Brousseau (1997) share this perspective. This view usually assumes a close relationship between the historical genesis and individual learning process, and supposes that students, with the assistance and guidance of a teacher are capable of constructing knowledge similar to that obtained historically by mathematicians. In particular, Lakatos (1976) demonstrated this view in his book, Proofs and Refutations, through an imaginary conversation between a teacher and pupils. The teacher and pupils support and criticize one another's claims from the perspective of various historical figures. However, the knowledge construction carried out by the teacher and pupils as presented by Lakatos is, in fact, the construction performed by prominent mathematicians including Euler, Legendre, and Cauchy. Lakatos' (1986) quasi-empirical view seems to ask students to learn mathematics by working like mathematicians (Chazan, 1990) prompting the question, "Is it also possible for elementary students to carry out knowledge constructions based on Euler's polyhedron theorem similar to those produced by mathematicians as discussed by Lakatos'?" In seeking a response to this question, this study focuses on (a) the knowledge constructions of mathematically gifted elementary students in comparison to those of mathematicians as discussed by Lakatos (1976), (b) how mathematically gifted fifth and sixth grade students justify Euler's polyhedron theorem, (c) the figures they suggest as counterexamples to Euler's polyhedron theorem, and (d) how they react when presented with counterexamples.

BACKGROUND

Literature Review

Sriraman found (2003) that the problem solving behaviors of mathematically gifted high school students' and those of non mathematically gifted students differed significantly. He reported that gifted students invest a considerable amount of time in trying to understand the problem situation, identifying the assumptions clearly, and devising a plan that was global in nature. Previous studies on the cognitive processes of mathematically gifted students have focused on generalization, abstraction, justification, and problem-solving (Krutetskii, 1976; Lee, 2005; Sriraman, 2003; 2004). Lee (2005) also found that mathematically gifted students have a tendency to advance to higher-level reasoning through reflective thinking.

Some researchers have analyzed the knowledge construction of students based on Lakatos' perspective (Athins, 1997; Boats, Dwyer, Laing, 2003; Borasi, 1992; Cox, 2004; Nunokawa, 1996; Reid, 2002; Sriraman, 2006). For example, Sriraman (2006) reconstructed the quasi-empirical approaches of six above average high school students' attempts to solve a counting problem and present the possibilities for mathematizing during classroom discourse in the spirit of Lakatos. Cox (2004) reported that the ability of high school students to proof improved after introducing them to the process of 'conjecture → proof → critique → accept or reject' in geometry classes. Borasi (1992) described the process where two high school students revised the definition of polygon and concluded that working on polygon "à la Lakatos" provided the context for valuable mathematical thinking and for activities that encourage participants to make use of their mathematical intuition and ability. Reid (2002) analyzed the problem-solving process of fifth-grade students and categorized their process of dealing with counterexamples based on monster-barring and exception-barring into three reasoning patterns. Athins (1997) reported that he observed a case of monster-barring on angles in a fourth grade mathematics class.

Euler's Polyhedron Theorem in Lakatos' Proofs and Refutations

In Lakatos' (1976) *Proofs and Refutations*, some justifications for Euler's theorem such as Cauchy's proof that appeared in the history of mathematics are shown in the dialogues between the teacher and pupils. For example, Lakatos has pupils Zeta and Sigma say the following explanation (pp. 70–72).

Step 1: For a polygon, $V = E$.
Step 2: For any polygon $V - E = 0$ (Figure 11.1a). If I fit another polygon to it (not necessarily in the same plane), the additional polygon has n_1 edges and n_1 vertices; now by fitting it to the original one along a chain of n'_1 edges and $n'_1 + 1$ vertices we shall increase the number of edges by $n_1 - n'_1$ and the number of vertices by $n_1 - (n'_1 + 1)$; that is, in the new 2-polygonal system there will be an excess in the number

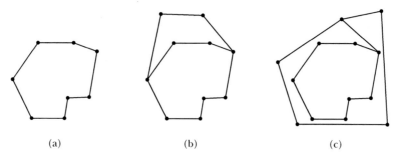

(a)   (b)   (c)

Figure 11.1

of edges over the number of vertices: $E - V = 1$; (Figure 11.1b); for an unusual but perfectly proper fitting see Figure 11.1c. "Fitting" a new face to the system will always increase this excess by one, or, for an F-polygonal system constructed in this way $E - V = F - 1$.

Step 3: I can easily extend my thought-experiment to 'closed' polygonal systems. Such closure can be accomplished by covering an open case-like polygonal system with a polygon-cover: fitting such a covering polygon will increase F by one without changing V or E. Or, for a closed polygonal system—or closed polyhedron—constructed in this way, $V - E + F = 2$.

Following the conjecture and proof, there appear counterexamples that refute the conjecture and proof. Lakatos called a counterexample that refutes lemma or subconjecture a *local* counterexample, and a counterexample that refutes the original conjecture itself a *global* counterexample (pp. 10–11). He suggested six types of counterexamples (Figures 11.2–11.7) which appeared in the history of mathematics as described below.

When a general counterexample is presented, there are five options. The first option considers the refuted conjecture incorrect and rejects it. The second option is to use the method of monster-barring in which the counterexample is seen as a monster, and the original conjecture is maintained (pp. 16–23). This method generates clearer definition, but it is not useful from a heuristic point of view because it does not improve the conjecture. The third option is the method of exception-barring in which the original conjecture is changed into a revised conjecture by adding a conditional clause that mentions an exception (pp. 24–27). This method does not guarantee that all exceptions are specified, and leaves the question of what is the range in which the theorem is valid. The fourth option is the method of monster-adjustment where the perspective under which the example was considered as a counterexample is seen as distorted, and the counterexample is interpreted as an example by readjusting the perspec-

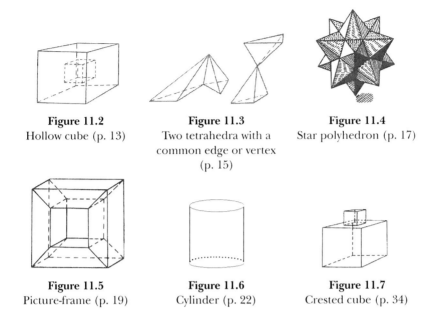

| **Figure 11.2** | **Figure 11.3** | **Figure 11.4** |
| Hollow cube (p. 13) | Two tetrahedra with a common edge or vertex (p. 15) | Star polyhedron (p. 17) |

| **Figure 11.5** | **Figure 11.6** | **Figure 11.7** |
| Picture-frame (p. 19) | Cylinder (p. 22) | Crested cube (p. 34) |

tive (pp. 30–33). The fifth option is the method of lemma-incorporation, where careful analysis of the proof is made to identify the guilty lemma. The lemma can then be incorporated in the conjecture to improve the refuted conjecture (pp. 33–42).

METHODOLOGY

Participants

Although there are diverse definitions of mathematical giftedness, there is no one universally accepted definition (e.g., Bluton, 1983; Miller, 1990; Gagne, 1991). In this study, Gagne's (1991) definition of mathematically gifted students as "students who are distinguished by experts to have excellent ability and potential for great achievements" was applied. Eleven fifth and sixth-grade male students (aged 10–12) from different Korean elementary schools in Gyeonggi province participated in the study. Five students were in the fifth-grade and six were in the sixth-grade. The sixth grade students were attending an advanced program for mathematically gifted students; three (A, B, and C) in a Korean government sponsored university program, and three (D, E, and F) in an office of education program. The fifth-grade students (G, H, I, J, and K), having passed a screening process which included a written test, an in-depth interview, and recommended from their school principal, were scheduled for admission to the university program. All students were motivated and confident of mathematics.

Tasks

The participants were presented with the following tasks:

Task 1: Explain what you know about the relationship between vertices (V), edges (E) and faces (F) in polyhedra. Explain how the relationship is justified.

Task 2: Is $V - E + F = 2$ true in all polyhedra? If not, when is it not true?

Task 3: If you consider a counterexample a polyhedron, how would you revise the theorem?

If you believe a counterexample is not a polyhedron, how would you revise the definition of a polyhedron?

Task 1 was designed to identify the participants' knowledge of the polyhedral theorem and to determine how they justify the theorem. Task 2 was developed to establish the types of counterexamples the participants identified. Task 3 was designed to observe how the participants resolved the disparity between the theorem and the counterexample.

The participants were familiar with the relationship between vertices, edges and faces, $V - E + F = 2$, before taking part in this study. However, they had not previously examined whether the theorem was true in all polyhedra, nor had they sought counterexamples to the theorem.

Data Collection and Analysis

This study was designed based on Yin's (2003) multiple case study methodology. The eleven participants were presented with the tasks in a set order and interviewed between November 2005 and January 2007. Each participant was video-taped by one researcher while they worked on the tasks and later while being interviewed by another researcher. The participants completed the tasks in approximately two hours. The video clips, transcriptions, observation reports and participants' worksheets were analyzed.

The analysis was conducted on three types of data collected: (a) the types of justification, (b) types of counterexamples, and (c) the methods for solving the conflict. The types of justification and counterexamples presented by the participants were analyzed using open coding (Strauss and Corbin, 1998). The types of justification were divided into two categories, and the counterexamples were categorized into four types, three of which were subdivided into two to three subtypes. The analysis of the participants' attempts to deal with the disparity between the counterexamples and the conjectures highlighted by the counterexamples was made using selective coding (Strauss and Corbin, 1998) which was based on "the method of monster-barring," "the method of exception-barring," "the method of monster-adjustment" and "the method of lemma-incorporation" suggested by Lakatos (1976). Cross-tabulation analysis was performed, and the results were examined by peers (Merriam, 1998).

RESULTS

Participants' Justification of Euler's Polyhedron Theorem

The participants' justification of the theorem can be divided in two ways; (a) to classify polyhedra into several categories and justify the theorem for each category of polyhedra, and (b) to attempt general justification without classifying polyhedra. The majority of participants justified the theorem by classifying polyhedra into categories and justifying the theorem. Participant D, in Episode 1 below, demonstrated this by logically explaining that the theorem is justified in prisms, pyramids, and prismoids:

Episode 1:

Participant D: First, in prisms, it seems to be justified in all cases.
Interviewer: Why is that?
Participant D: (Drawing figures) Well, look at an n-angle prism. A rectangular prism, it's called that because the bases are rectangles. So, there are four vertices on the top face and four on the bottom face, so, the number of vertices is $2n$. Also, the number of edges is $3n$ because there are four edges on the top face, four on the bottom face, and four on the lateral sides. And, the number of faces is $n + 2$ because there are four faces on the lateral sides plus the top and bottom faces. In the case of a pentagonal prism, also, the number of faces is $n + 2$, as there are five lateral faces plus the bases (top and bottom faces). "$V - E + F$" stands for "number of vertices – number of edges + number of faces," and in n-angle prisms, it is "$2n - 3n + (n + 2)$," so "$V - E + F$" equals 2.
Interviewer: Yes.

Participant D: So, I'm done with prisms... in pyramids, too, it is justified all the time.

Interviewer: Please explain.

Participant D: ...an n-angle pyramid. It's justified because the number of its vertices is $n + 1$, and it has $2n$ edges and $n + 1$ faces. If you add the number of vertices and the number of faces, and then subtract the number of edges, you get 2.

Participant D provided explanations using polyhedra such as rectangular prism in the case of prisms, pyramids, and prismoids. Rectangular prism is a generic example (Mason & Pimm, 1984) which represents general n-angle prism. In the case of regular polyhedra or a polyhedron like the soccer ball, D investigate the theorem application by counting the numbers of points, edges, and faces of specific solids.

Participant B did not categorize solid figures but instead attempted generalized justification. He started with a point (see Figure 11.8) and verified $V - E + F$ as the number of points, lines, and faces gradually increased. According to him there is only one V at first, but V and E or E and F increases by 1 respectively as procession is made from (a) to (g) and, $V - E + F$ is maintained at 1. In the last stage, when one face is covered in (g), he proved that $V - E + F = 2$, based on the fact that the number of F increases by 1. This justification is similar to the explanation of pupils Zeta and Sigma in Lakatos (1976, pp. 70–72).

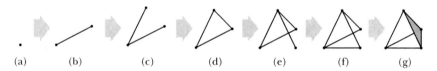

(a)　　　(b)　　　(c)　　　(d)　　　(e)　　　(f)　　　(g)

Figure 11.8

After justifying Euler's theorem, all the participants expressed the view that there might be a polyhedron with which Euler's theorem was not true. For example, participant D, as indicated in Episode 2, thought that the theorem would not hold in all polyhedra.

Episode 2:

Participant D: Well ... first, it is justified in regular polyhedra without exception, because there are only five kinds of regular polyhedra. I think it is justified in all of the five, and then, it is justified, first, in prisms and pyramids. So, I think it is justified in the majority of general polyhedra ...

Interviewer: Then, do you think there are some cases in which it doesn't apply?

Participant D: *In some cases... I think it won't apply in all cases.* (Starts drawing figures to find solids with which the polyhedral theorem is not true)

Although participant B justified the theorem using a general method, he tried to find a counterexample, thinking that there still might be one. All the participants express the view that there had to be an example in which the theorem does not apply.

Solid Figures Suggested by Participants as Counterexamples

Participants suggested various types of solid figures as counterexamples to the theorem. The solid figures suggested by the participants were categorized into the four groups below.

Solids with Curved Surfaces

Six participants (B, C, E, F, H and I) suggested solids with curved surfaces such as a cone (Figure 11.9), a cylinder (Figure 11.10) and a sphere (Figure 11.11) as counterexamples. Each participant had a different reason for suggesting the cone as a counterexample. Participant F drew the net of a cone in order to count the points, lines, and faces. He claimed that the circle in the net was not counted as an edge because it was a curve, but the radius of the sector had to be counted as an edge because it was a straight line ($V = 1$, $E = 2$, $F = 2$, $V - E + F = 1$). Participant E insisted that the radius of the sector in the net could not be counted as an edge because it was not actually seen in the solid, and thus, $V = 1$, $E = 0$, $F = 2$, $V - E + F = 3$. Participant H said that a cone provided a counterexample, "Because you can't say how many edges there are in a circle."

Figure 11.9   Cone   **Figure 11.10**   Cylinder   **Figure 11.11**   Sphere

Solids Made of Multiple Polyhedron Sharing Points, Lines, or Faces

Nine participants (A, B, C, D, E, F, G, H and I) cited solids made of two polyhedra sharing points, lines, and faces as counterexamples. These solids can be divided into (a) solids that completely share some points, lines, or faces (Figure 11.12 through Figure 11.15), and (b) solids that only partially share lines or faces (Figure 11.16 through Figure 11.19).

Solids that Completely Share Points, Lines, or Faces

In solids that share one point as shown in Figure 11.12, the theorem holds in each polyhedron and two polyhedra share a point, $V - E + F = 3$. Participants also suggested solids that share an edge (Figure 11.13) and those that share a face completely (Figure 11.14 or Figure 11.15) are counterexamples.

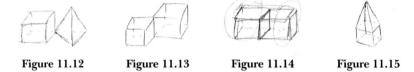

| Figure 11.12 | Figure 11.13 | Figure 11.14 | Figure 11.15 |

Solids that Partially Share Lines or Faces

Solids such as in Figure 11.14 and Figure 11.15 raised the issue with participants of whether it is appropriate to consider shared faces as separated faces. Participants suggested that modified solids that partially share lines or faces were counterexamples.

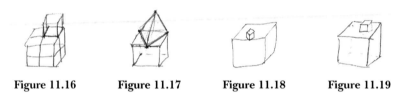

| Figure 11.16 | Figure 11.17 | Figure 11.18 | Figure 11.19 |

Where edges are partially shared (Figure 11.16) and where an edge is divided (Figure 11.17) the participants reflected on how to count the number of edges. And a counterexample such as Figure 11.19 led the participants to contemplate the question, "Is it appropriate to consider the face created by joining two faces a face?" Lakatos (1976, p. 74) called this a "ring-shaped face" (Figure 11.20).

Figure 11.20   Ring-shape face

Polyhedra with Holes

The third type of solids that the participants (A, B, C, G, J and K) suggested as counterexamples is solids with holes as shown in Figure 11.21 through Figure 11.23. These counterexamples also prompted the participants to rethink the definition of face.

Figure 11.21

Figure 11.22

Figure 11.23

Polyhedra Containing Other Polyhedra

Eight participants (A, B, C, D, F, G, J and K) also suggested solids that are polyhedra containing other polyhedra are counterexamples. Counterexamples of this type can be subdivided into three subtypes. The first is the type in which other solids—not sharing any face, point or line—exist in certain solids (Figure 11.24). The second is the type in which two solids completely share a face (Figure 11.25). The third type is one in which a figure exists inside another and the two figures share part of a face.

Figure 11.24

Figure 11.25

Figure 11.26

Participants' Responses to the Disparity Created by Counterexamples

The participants' responses to the disparity between counterexamples and the theorem are divided into four categories; the method of monster-barring, the method of exception barring; the method of monster adjustments and new conjectures.

The Method of Monster-barring

Participants D and E used the method of monster-barring. In Episode 3, participant E, suggested cones, cylinders, and spheres as counterexamples, and wondered how to determine the numbers of points, lines, and faces in these figures. He then stated, "*A polyhedron is a solid figure made of multiple polygons*", and that the curved surface is not a polygon, and thus, solids with curved surfaces are not polyhedra but monsters.

Episode 3:

Participant E: Cones have curved surfaces, so I think they will not work.
Interviewer: What's wrong with curved surfaces?

Participant E: Because in a curved surface, you can't count the number of edges, and faces... Can you count the number of
faces? But the number of vertices is one... I think there is
no edge, in the definition that I think of.

Interviewer: Can you say a cone is a polyhedron? Euler's theorem is
about polyhedra.

Participant E: When you talk about curved surfaces, a sphere has a
curved surface, and a sphere has one face,... but no dis-
tinguishable edge or point, I guess there are none.

Interviewer: What do you think is the definition of a polyhedron?

Participant E: I think it is made of faces that have angles. (Writing down
the definition) "Polyhedron = solid figure made of mul-
tiple polygons"

Participant D also used the method of monster-barring where polyhedra existed in other polyhedra. He used the method of monster-barring stating that a
polyhedron signified "one" solid figure, and that the polyhedron in which there
is another polyhedron meant two different solid figures. He modified the definition of polyhedron as "one solid figure surrounded by multiple polygons."

The Method of Exception-Barring

Participants A, D, F, G, and I were observed attempting the method of
exception-barring. Participant F defined a polyhedron as a "figure made
of faces." So the solid figures with curved surfaces are polyhedra because
curved surfaces are faces. To exclude cones, cylinders, and spheres as exceptions, participant F modified the original conjecture to "*In all polyhedra
excluding those made of curved faces, $V - E + F = 2$.*" Participant I, Episode 4,
also used the method of exception-barring by modifying the theorem to "*In
polyhedra that do not include a circle, $V - E + F = 2$.*"

Episode 4:

Interviewer: (Pointing to the sphere and cylinder.) Then, can we call
them polyhedra, too?

Participant I: It has one or more faces... We can call them polyhedra.

Interviewer: Then, don't we need to modify this ($V - E + F = 2$)?

Participant I: Yeah...

Interviewer: How can we change it?

Participant I: (Thinking hard) So, if a circle is included... I guess only
the polyhedra without any circles belong to this category
($V - E + F = 2$), don't they?

Participant I suggested two rectangular solids that share one edge (Figure 11.27) are another counterexample. Then he redefined the theorem

to "In polyhedra that do not include a circle and are not attached to other polyhedra, $V - E + F = 2$." Participant G found solid figures with holes as counterexamples, and modified the theorem to "In polyhedra which are not completely penetrated by a hole, $V - E + F = 2$."

Figure 11.27

The Method of Monster-Adjustment

Participants B, D, E, F, and G tried the method of monster-adjustment to convert a counterexample into an example. Participant B thought, after finding the counterexample in which part of a face was shared by two figures, that the justification of Euler's theorem depended on whether to consider the edge divided by a point as one or two.

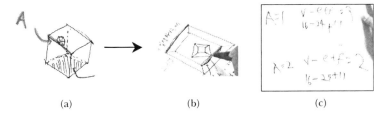

(a)   (b)   (c)

Figure 11.28

Participant B compared the results when the edge (line A in Figure 11.28a) divided by a point was counted as one (A = 1) and when it was counted as two (A = 2). Then, B, Episode 5, explained the reason why the edge divided by a point in this solid figure should be counted as two.

Episode 5:

Interviewer: Is the solid a polyhedron?
Participant B: It is a polyhedron.
Interviewer: Then, what can we do?
(Participant B is writing)
Participant B: If there is a vertex in the middle of an edge (even when it is not on the exact center), the left and right sides of the vertex should be separately counted... It is absolutely necessary to separately count this part (left part of line A) and that part (right part of line A). In the case of plane figures, we count any line between two points separately... In solids, to make it (the value of $V - E + F$) become 2, you need to count the left and right sides of the point separately.

Two polyhedra that completely share a face, with one inside the other (Figure 11.25) were also considered not to be a counterexample by one of the participants after the method of monster-adjustment was used. Participant D claimed that the figure was not a counterexample because it was considered a sunken solid without a lid, rather than two solids sharing one face.

For the ring-shaped face (Figure 11.20), some participants preferred to use the method of monster-barring by not considering it as a face, and subsequently, employed the method of monster-adjustment by not considering the solid figures with ring-shaped faces as counterexamples (e.g., $V = 16$, $E = 24$, $F = 10$, and thus, $V - E + F = 2$ in Figure 11.19). Participant I, who used the method of exception-barring for cylinders and spheres, used the method of monster-adjustment for the cone, considering the polyhedral theorem to be justified under the condition of $V = 1$, $E = 1$, and $F = 2$.

New Conjectures

The participants' approaches were not limited to monster-barring, exception-barring, and monster-adjustment which are similar in the sense that they are used to support the formula of $V - E + F = 2$. Participants also suggested two types of new conjectures. One involved the participants searching for a new formula about the value of $V - E + F$ with which to express the relationships between the points, lines, and faces in solid figures including the counterexamples they found. New conjectures suggested by participants are summarized in Table 11.1.

TABLE 11.1 Summary of Participants Conjectures

$V - E + F$	Conditions	Participants
0	If the ring-shaped face is not a face, in polyhedra with hole(s)	G
1	In polyhedra including a circle	I
3	In polyhedra that completely share either a point or a line with other polyhedra	E and F
	If solid figures are attached at a vertex, edge, or face	H
4	In polyhedra which contained other polyhedra such as a hollow cube	F

The other type of conjecture, suggested by participant A, relates to the necessity of considering new elements other than points, lines, and faces. He proposed, Episode 6, that a formula that including three-dimensional elements be developed.

Episode 6:

Participant A: In the two-dimensional circumstance, a rule can be easily found using just V, E, and F, but in the three dimensions,

a new element of space is added. So, if Euler's theorem is
a formula established using two-dimensional elements, I
guess we can make a new formula that exclusively applies
to the third dimension including space, can't we?

Interviewer: The new element of three dimensions. Can we really do it
if we consider that?

Participant A: Yes, I think so.

Interviewer: Then how can we determine the numbers in the three
dimensions?

Participant A: Space.

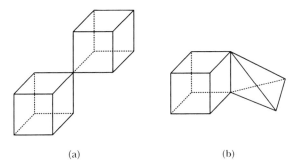

(a)   (b)

Figure 11.29

After that, $V - E + F - S = 1$ and $V - E + F + S = 3$ were proposed as new conjectures, and it was confirmed that $V - E + F - S = 1$ is justified with the type
of solid figures in Figure 11.29a: $V = 15$, $E = 24$, $F = 12$, $S = 2$, $V - E + F - S = 1$;
Figure 11.29b: $V = 10$, $E = 17$, $F = 10$, $S = 2$, $V - E + F - S = 1$). This conjecture
led participant A to think that the polyhedral theorem could be expanded
to four-dimensional solids.

DISCUSSION

Polyhedra, which the participants studied prior to the research, were limited to the category of regular polyhedra, prisms, pyramids, prismoids, and
semi-regular polyhedra such as soccer balls, all satisfied Euler's theorem.
Nevertheless, the participants thought that there must be some polyhedra
for which the theorem was not valid. This belief appears to resulted from
the method of justification that the majority of participants used. The value
of $V - E + F$ can be obtained by counting the numbers of points, edges, and
faces in the case of prisms, pyramids, and prismoids (e.g., in n-angle prism,
$V = 2n$, $E = 3n$, $V = n + 2$, and thus, $V - E + F = 2$). However, this justification
fails to provide information about new kinds of solids that have yet to pass
this test. The participants' view that there must be polyhedra with which

the polyhedral theorem was not valid indicates they belive that the scope of polyhedra is extensive. This view is supported by the various types of solids that the participants presented as counterexamples.

A strong similarity exists between solid figures suggested by the participants as counterexamples and those discussed by Lakatos (1976). The first type of counterexample that participants found, solids with curved surfaces, appeared as cylinders in Lakatos (p. 22). The second type, two or more polyhedra that shares points, lines, or faces, was discovered by mathematicians Hessel (figures that share points or lines) and Lhulier (cube with crest) in 1832 and 1813, respectively (pp. 15, 34). The third type of counterexample was first discovered by Lhuilier (p. 19). In addition to the tunnel and picture frame mentioned in Lakatos, participants also found a polyhedron which is not completely penetrated. The fourth type, polyhedra within polyhedra, was discovered by Lhulier and Hessel based on the idea obtained by observing the crystalloid of mineralogic collection enclosed within a translucent crystalloid (p. 13).

Counterexamples can be used to help students develop their mathematical reasoning (Lakatos, 1976; Boats, et al., 2003). In this study participants examined concepts such as polyhedron and face and created new definitions. The counterexamples discovered by participants also encouraged them to examine more closely the definition of terms. The ring-shaped face in particular prompted some participants to reconsider the definition of polygon. They asserted that it could not be called a polygon, because the figure did not comply with the sum of interior angles of n-polygon $180 \times (n-2)$. This suggests that the formula for the sum of interior angles of a polygon was seen as a definitive property that determines whether the figure was a polygon or not. This method of defining a polygon is similar to the definition of polyhedron stated by Baltzer (Lakatos, 1976, p.16), "polygon system with which the equation of $V - E + F = 2$"

In Reid (2002) and Athins (1997), the method of monster-barring and the method of exception-barring were observed among elementary students. The method of monster-barring, the method of exception-barring, the method of monster-adjustment, and new conjectures were observed among the participants in this study. The participants did not reject the original theorem and attempted to develop new conjectures that comprised counterexamples, and of the five participants who used the method of exception-barring, four developed new conjectures. In the past, there have been cases in which counterexamples were first recognized as monsters and excluded, but later reintroduced and accepted as examples (e.g. Lakatos, 1976, p.31). This ability to review and change a position was also demonstrated by the participants. Initially, they used the monster-barring method or exception-barring method for the counterexamples they identified, but they attempted to include the counterexamples within the scope of examples through monster adjustment or new conjecture. Krutetskii, (1976) Sriraman, (2004) point out that this flexibility of thinking is an attribute of mathematically gifted.

Lakatos (1976) argues that the method of lemma-incorporation is a productive way of refining conjecture based on the proof. Proof-analysis is a prerequisite to this method and, as Nunokawa points out (1996) proof-analysis is an important component of proofs and refutations. However, in this study, the method of lemma-incorporation and proof-analysis was not observed. When participant B, provided proof of increasing the elements of polyhedra, was encouraged considering the validity of his proof for a counterexample (Figure 11.18), he provided a monster adjustment solution stating, "It's not the proof that's wrong, but there is a problem with this solid."

CONCLUSION

This study focuses on the constructions of mathematically gifted fifth or sixth-grade students in solving tasks related to Euler's polyhedron theorem and compares them to those of mathematicians discussed by Lakatos (1976). By analyzing ninth grader students notion of proof, Sriraman (2004) reports that the processes used by gifted students demonstrate remarkable isomorphism to those employed by professional mathematicians, This study also shows parallels in constructions of mathematically gifted fifth and sixth grade student and mathematicians discussed by Lakatos. With the exception of the method of lemma incorporation and proof-analysis, counterexamples and the method for solving conflicts between the theorem and counterexamples suggested by the participants demonstrated remarkable similarities to those presented in the history of mathematics.

REFERENCES

Athins, S. (1997). Lakatos' proofs and refutations comes alive in an elementary classroom. *School Science and Mathematics, 97*(3), 150–154.

Bluton, C. (1983). Science talent: The elusive gift. *School Science and Mathematics, 83*(8), 654–664.

Boats, J. J., Dwyer, N. K., Laing, S., & Fratella, M. P. (2003). Geometric conjectures: The importance of counterexamples. *Mathematics Teaching in the Middle School, 9*(4), 210–215.

Borasi, R. (1992). *Learning mathematics through inquiry.* Portsmouth, NH: Heinemann.

Branford, B. (1908). *A study of mathematical education.* Oxford: Clarendon Press.

Brousseau, G. (1997). *Theory of didactical situations in mathematics* (N. Balacheff, M. Cooper, R. Sutherland & V. Warfield , Ed. and Trans.). Dordrecht: Kluwer Academic.

Chazan, D. (1990). Quasi-empirical views of mathematics and mathematics teaching. *Interchange, 20*(1), 14–23.

Clairaut, A. C. (1741). *Éléments de géométrie.* Paris: Gauthier-Villars.

Clairaut, A. C. (1746). *Éléments de algébre.* Paris: Rue Saint Jacques.

Cox, R. (2004). Using conjectures to teach sutdents the role of proof. *Mathematics Teacher, 97*(1), 48–52.

Freudenthal, H. (1983). *Didactical phenomenology of mathematical structures.* Dordrecht: D. Reidel Publishing Company.

Freudenthal, H. (1991). *Revisiting mathematics education.* Dordrecht: Kluwer Academic.

Gagne, F. (1991). Toward a differentiated model of gifted and talent. In N. Colangelo & G.A. Davis (Eds.), *Handbook of gifted education* (pp. 65–80). Boston: Allyn and Bacon.

Klein, F. (1948). *Elementary mathematics from an advanced standpoint: Arithmatic, algebra, analysis.* (E. R. Hedrick & C. A. Noble, Trans.). New York: Dover. (Original work published 1924).

Krutetskii, V. A. (1976). *The psychology of mathematical abilities in school children.* Chicago: The University of Chicago Press.

Lakatos, I. (1976). *Proofs and refutations: The logic of mathematical discovery.* Cambridge: Cambridge University Press.

Lakatos, I. (1986). A renaissance of empiricism in the recent philosophy of mathematics? In T. Tymoczko (Ed.), *New directions in the philosophy of mathematics* (pp. 29–48). Boston: Birkhauser.

Lee, K. H. (2005). Mathematically gifted students' geometrical reasoning and informal proof. In L. C. Helen, & L. V. Jill (Eds.), *Proceedings of the 29th Conference of the International Group for the Psychology of Mathematics Education* (Vol 3. pp. 241–248), Melbourn, Australia: PME.

Mason, J., & Pimm, D. (1984). Generic examples: Seeing the general in the particular. *Educational Studies in Mathematics, 15,* 277–289.

Merriam, S. B. (1998). *Qualitative research and case study applications in education.* John Wiley & Sons, Inc.

Miller, R. C. (1990). *Discovering mathematical talent.* (ERIC Digest No. E482).

Nunokawa, K. (1996). Applying Lakatos' theory to the theory of mathematical problem solving. *Educational Studies in Mathematics, 31,* 269–293.

Reid, D. (2002). Conjectures and refutations in grade 5 mathematics. *Journal for Research in Mathematics Education, 33*(1), 5–29.

Sriraman, B. (2003). Mathematical giftedness, problem solving, and the ability to formulate generalizations: The problem-solving experiences of four gifted students. *Journal of Secondary Gifted Education, 14*(3), 151–165.

Sriraman, B. (2004). Gifted ninth graders' notions of proof: Investigating parallels in approaches of mathematically gifted students and professional mathematicians. *Journal for the Education of the Gifted, 27*(4), 267–292.

Sriraman, B. (2006). An ode to Imre Lakatos: Quasi-thought experiments to bridge the ideal and actual mathematics classrooms. *Interchange, 37*(1–2), 151–178.

Strauss, A., & Corbin, J. (1998). *Basics of qualitative research* (2nd ed.). Thousand Oaks, CA: Sage Publications.

Toeplitz, O. (1963). *The calculus—A genetic approach* (L. Lange, Trans.). Chicago: The University of Chicago Press. (Original work published 1949).

Yin, R. K. (2003). *Case study research: Design and methods* (3rd ed.). Thousand Oaks, CA: Sage Publications.

ACKNOWLEDGMENT

Reprint of Yim, J., Song, S., & Kim, J. (2008). The mathematically gifted elementary students' revisiting of Euler's polyhedron theorem. *The Montana Mathematics Enthusiast*, 5(1), 125–142. Reprinted with permission from The Montana Mathematics Enthusiast and Information Age Publishing. ©2008 Information Age Publishing.

CHAPTER 12

MATHEMATICALLY PROMISING STUDENTS FROM THE SPACE AGE TO THE INFORMATION AGE

Linda Jensen Sheffield
Northern Kentucky University

THE SPACE AGE

On October 4, 1957, with the launch of Sputnik 1 by the Soviet Union, the world entered the Space Age and the United States became quite concerned that the Soviet Union had a head start in the space race. A year later, realizing that the support of gifted and talented mathematics and science students was critical to national security, the United States federal government passed the National Defense Education Act (NDEA), providing aid to education in the United States at all levels, primarily to stimulate the advancement of education in science, mathematics, and modern foreign languages. Also, during this time, "new math" was introduced with an emphasis on more abstract concepts and unifying ideas. One of the most unique of the projects developed during that time, the Comprehensive School Mathematics Program (CSMP) from McREL, Mid-continent Research for

Education and Learning, continues to be available online at http://ceure. buffalostate.edu/~csmp/. Although never fully implemented as intended, some of the "new math" projects along with the NDEA contributed to the dominance of the United States in science and technology in the latter part of the twentieth century as they inspired thousands of students to enjoy mathematical investigations and to pursue degrees in mathematics, science and technology.

On July 16, 1969, the Apollo 11 launched from the Kennedy Space Center and on July 20, 1969, Commander Neil Armstrong became the first man on the moon and said the historic words, "One small step for man, one giant leap for mankind." The sixth and final manned moon landing occurred in December 1972, and the United States declared victory in the space race. For fifteen years, Americans had supported gifted and talented students interested in learning mathematics and science, especially as related to space technology, but what has happened since that time?

THE GROWTH OF TECHNOLOGY

Partially in reaction to the "new math," the 1970s saw a strong "back-to-basics" movement with an emphasis on basic skills such as computation. In 1980, the National Council of Teachers of Mathematics (NCTM) published *An Agenda for Action* noting that the most important basic skill was problem solving. The following statement, from this same report pointed to the growing recognition of the importance of the development of gifted mathematics students:

> The student most neglected, in terms of realizing full potential, is the gifted student of mathematics. Outstanding mathematical ability is a precious societal resource, sorely needed to maintain leadership in a technological world. (NCTM, 1980, p. 18)

In 1983, The National Commission on Excellence in Education warned in its report, *A Nation at Risk*, that the skills and knowledge of the U.S. workforce would have to improve dramatically in order for the nation to remain internationally competitive. In 1989, President George H. W. Bush convened an Education Summit with the nation's Governors and adopted six National Education Goals. The fifth goal was: "U.S. students will be first in the world in mathematics and science achievement by the year 2000." In spite of the public acknowledgement of the importance of students with high-level skills in mathematics and science, little has been done in the past 25 years to support our most promising students.

THE INFORMATION AGE

In 1993, Richard Riley, the U. S. Secretary of Education, in the introduction to National Excellence: A Case for Developing America's Talent, stated, "All of our students, including the most able, can learn more than we now expect. But it will take a major national commitment for this to occur." (Ross, 1993, p. iii) The report goes on to point to a "quiet crisis in educating talented students" with the following statement: "The United States is squandering one of its most precious resources—the gifts, talents, and high interests of many of its students" (Ross, 1993, p. 1).

The year after this report came out, the NCTM appointed a Task Force on Mathematically Promising Students to analyze this issue specifically for mathematics. The Task Force agreed that a major national commitment was needed to turn around this quiet crisis for mathematically promising students who were defined as "those who have the potential to become the leaders and problem solvers of the future." The Task Force called for a strategy that seeks to greatly increase the numbers and levels of mathematically promising students by maximizing their ability, motivation, beliefs, and experiences/opportunities. The report pointed out that these four factors are all variables that could and should be increased with proper support and encouragement. Noting research on brain functioning that demonstrates that significant changes in the brain are due to experiences, the report called on administrators, teachers, parents and students themselves to make sure that all students have the opportunity to experience the joy of solving challenging mathematical problems on a regular basis and that high-level mathematics courses are available to all students regardless of where they go to school. Recognizing that the culture in the United States often works against students' desire to excel in science, technology and mathematics, the report also noted the importance of students' realizing that excellence in mathematics is not only possible, but also leads to careers in fulfilling and intriguing areas. (Sheffield, et al, 1995) The recent popularity of the television series *NUMB3RS* goes a long way toward supporting this goal, but much more is needed.

THE TWENTY-FIRST CENTURY

By 2000, it was evident that the United States was a long way from the goal of being first in the world in math and science. The Trends in International Mathematics and Science Study (TIMSS) in 1995 and the repeat of the study in 1999 and 2003 showed that not only were we not first, but top students in the United States were not at the same level as top students in other countries. In1995, 9% of U. S. fourth graders and 39% of Singapore fourth graders scored above the 90th percentile on the mathematics portion of the TIMSS test. That year, 5% of U. S. eighth graders and 45% of Singapore eighth graders scored above the 90th percentile on the TIMSS mathematics test. By 2003, 40% of the eighth grade students in Singapore, 38% of eighth graders in Taiwan, and 7% of U. S. eighth graders scored at the most advanced level. Although this was an improvement for students in the United States, it was still far behind other developed countries.

Similar results were found by the Program for International Student Assessment (PISA). In 2003, U.S. performance in mathematics literacy and problem solving was lower than the average performance for most OECD (Organization for Economic Co-operation and Development) countries. Even the highest U.S. achievers (those in the top 10 percent in the United States) were outperformed on average by their OECD counterparts (National Center for Education Statistics, 2003).

The *No Child Left Behind Act of 2001* had as a major purpose that all students reach proficiency on challenging state standards and assessments, closing the achievement gap between high and low-achieving students. But what happens to students for whom moving toward proficiency is moving backwards when there is a goal to close the achievement gap between high and low-performing students?

In a study of the effects of teachers and schools on student learning, William Sanders and his staff at the Tennessee Value-Added Assessment System put in this way: "Student achievement level was the second most important predictor of student learning. The higher the achievement level, the less growth a student was likely to have" (DeLacy, 2004, p. 40).

Certainly one way to close the achievement gap between high and low-performing students is to slow down the learning of high-performing students, but is that a goal that we can afford?:

> The United States is losing its edge in innovation and is watching the erosion of its capacity to create new scientific and technological breakthroughs....If America is to sustain its international competitiveness, its national security and the quality of life of its citizens, then it must move quickly to achieve significant improvements in the participation of all students in mathematics and science. (Business-Higher Education Forum, 2005, p. 1, 3)

In 2005, the Annual Conference of the National Association of Gifted Children (NAGC) featured a special strand on Mathematics and Science with a keynote address by Jim Rubillo, the Executive Director of the National Council of Teachers of Mathematics and Gerry Wheeler, the Executive Director of the National Science Teachers Association, and NAGC appointed a Math/Science Task Force to continue this work. If the United States is to maintain leadership in this technological world, it is critical that we collaborate to take immediate drastic action to recognize, support, create and develop the mathematical promise in large numbers of students and their teachers—male and female; black and white; preschool through graduate school; rich and poor; rural and urban. As we approach the fiftieth anniversary of Sputnik and the National Defense Education Act, let's join together to inspire a new generation of students to excel in these areas critical to the welfare of our country and indeed of the entire world.

REFERENCES

Achieve, Inc. and the National Governor's Association. (2005). *An action agenda for improving America's high schools: 2005 national education summit on high schools.* Retrieved July 31, 2005, from http://www.nga.org/Files/pdf/0502actionagenda. pdf.

Adelman, C. (1999). *Answers in the tool box: Academic intensity, attendance patterns, and Bachelor's degree attainment.* Retrieved February 15, 2005, from http://www. ed.gov/pubs/Toolbox/index.html.

Anderson, S. (Summer 2004). *The multiplier effect. International education.* National Foundation for American Policy. Retrieved February 15, 2005, from http:// www.nfap.net/.

Association of American Universities. (2005). A national defense education act for the 21st century: Renewing our commitment to the U. S. students, science, scholarship, and security. Retrieved December 13, 2005 from http://www. aau.edu/education/NDEAOP.pdf#search='National%20Defense%20Education%20Act'

Business-Higher Education Forum. (January 2005). *A commitment to America's future: Responding to the crisis in mathematics and science education.* Retrieved July 31, 2005 from http://www.bhef.com/MathEduReport-press.pdf.

Colvin, G. (July 25, 2005). America isn't ready: Here's what to do about it. *Fortune, 152*(2), 70–82.

DeLacy, M. (June 23, 2004). The 'No Child' law's biggest victims? An answer that may surprise, *Education Week, 23*(41), 40.

Florida, R. (2005). *The flight of the creative class: The new global competition for talent.* New York: Harper Business.

Friedman, T. L. (2005) *The world is flat: A brief history of the twenty-first century.* New York: Ferrar, Straus, and Giroux.

Giambrone, T. M. *Comprehensive school mathematics preservation project.* Retrieved December 14, 2005 from http://ceure.buffalostate.edu/~csmp/.

Lewis, J. A. (October 2005). *Waiting for Sputnik: Basic research and strategic competition.* Retrieved December 14, 2005, from http://www.csis.org/media/csis/ pubs/051028_waiting_for_sputnik.pdf

National Center for Education Statistics. (December 2000). *Pursuing excellence: Comparisons of International eighth-grade mathematics and science achievement from a U.S. perspective, 1995 and 1999.* Retrieved July 31, 2005 from http://nces. ed.gov/pubsearch/pubsinfo.asp?pubid=2001028.

National Center for Education Statistics. (2003). *Program for international student assessment (PISA) 2003 summary.* Retrieved December 10, 2005 from http://nces.ed.gov/surveys/pisa/PISA2003Highlights.asp.

National Commission on Educational Excellence. (April 1983). *A nation at risk: The imperative for education reform.* Retrieved May 25, 2005, from http://www.ed.gov/pubs/NatAtRisk/index.html.

National Council of Teachers of Mathematics (NCTM). (1980). *An agenda for action: Recommendations for school mathematics of the 1980s,* Reston, VA: NCTM.

National Education Goals Panel. (1990). *Building a nation of learners.* Retrieved June 10, 2005, from http://govinfo.library.unt.edu/negp/.

National Science Board. (2004). *An emerging and critical problem of the science and engineering labor force.* Retrieved March 13, 2005, from http://www.nsf.gov/sbe/srs/nsb0407/start.htm.

Public Law 107-110 (January 8, 2002) *The elementary and secondary education act (The no child left behind act of 2001).* Retrieved May 30, 2005, from http://www.ed.gov/policy/elsec/leg/esea02/index.html.

Ross, Pat O'Connell (Project Director). (1993). *National excellence: A case for developing America's talent.* Washington, DC: U.S. Department of Education, Office of Educational Research and Development.

Sheffield, L. J. (Fall, 2005) Mathematics: The pump we need to combat the brain drain. *Gifted Education Communicator, 36*(3).

Sheffield, L. (chair), Bennett, J., Berriozábal, M., DeArmond, M., and Wertheimer, R. (December 1995) *Report of the task force on the mathematically promising.* Reston, VA: NCTM News Bulletin, Volume 32.

Task Force on the Future of American Innovation. (February 16, 2005). *The knowledge economy: Is America losing its competitive edge? Benchmarks of our innovation future.* Retrieved July 31, 2005, from http://www.futureofinnovation.org/PDF/Benchmarks.pdf.

Trends in International Mathematics and Science Study (TIMSS). (2004). *TIMMS 2003 results.* Retrieved March 23, 2005 from http://nces.ed.gov/timss/Results03.asp.

Trends in International Mathematics and Science Study (TIMSS). (2000). *TIMMS 1999 results.* Retrieved March 23, 2005 from http://nces.ed.gov/timss/Results.asp.

United States Commission on National Security/21st Century. (February 15, 2001). *Roadmap for national security: Imperative for change.* Retrieved July 31, 2005, from http://www.au.af.mil/au/awc/awcgate/nssg/phaseIIIfr.pdf

ACKNOWLEGMENT

Reprint of Sheffield, L. (2005). Mathematically promising students from the space age to the information age. *The Montana Mathematics Enthusiast, 3*(1), 104–109. Reprinted with permission from The Montana Mathematics Enthusiast. ©2005 Linda Jensen Sheffield.

CHAPTER 13

REVISITING THE NEEDS OF THE GIFTED MATHEMATICS STUDENTS

Are Students Surviving or Thriving?

Alan Zollman
Northern Illinois University

ABSTRACT

Are gifted students surviving or thriving? In the mathematics classroom, too often, the answer is "it depends." The public school has priorities other than the individual gifted student's needs. The United States—indeed the modern world—demands all students, particularly the highly and extremely gifted, to attain self-actualization. This paper (a) reviews the different types of gifted students, (b) delineates different approaches to fulfilling the needs of the gifted, (c) discusses the current barriers to meeting the needs of the highly and extremely gifted, (d) examines the opportunities available for the gifted, and, (e) speculates on the future for gifted mathematics student.

Creativity, Giftedness, and Talent Development in Mathematics, pages 277–286
Copyright © 2008 by Information Age Publishing

INTRODUCTION

James is a 10-year-old sixth grader. He previously skipped the second grade and fifth grade math. This week he was advanced to a ninth-grade algebra class, his third new math course of the semester, because of his advanced math talent. This temporary solution is acceptable to James. He was bored and frustrated with the other math courses. But he will complete all of the high school math offerings by his ninth grade. What will he do in high school—again be bored and frustrated with the material he already knows? Will the school, indirectly and unknowingly, teach James to become a "gifted underachiever" (House, 1987) with hostility towards school, poor study habits, and low levels of self-discipline?

Instead of going into a specialized gifted program that considers his emotional, social, and intellectual needs, James is academically advanced into higher-level classes. It is the "easiest and most economical" plan from the school's perspective. No specialized teachers or classes are required. Is James being helped or harmed by this approach? Is James being allowed to thrive mathematically? Where is the Individualized Educational Plan (IEP) specific designed for James's emotional needs, social requirements, personal desires, and individual abilities?

The school wants to keep James in the sixth grade. Why? The teacher says the school wants his math scores on the state assessment compiled with their other sixth graders' scores. The school is concerned with the high-stakes state assessment. Is this survival tactic of the school best for James' needs?

In 1991 the U.S. Department of Education issued *AMERICA 2000: An Education Strategy*. Its fourth national education goal was "by the year 2000, U.S. students will be first in the world in science and mathematics achievement" (p. 9). The U.S. did not meet this goal. According to the Third International Math and Science Survey (TIMSS, 1999) and Trends in International Math and Science Study (TIMSS, 2003), the U. S. has not improved its international "ranking" since 1991.

Recently, in 2002 the U.S. Department of Education implemented the *No Child Left Behind (NCLB) Act of 2001*. This federal law requires states to collect and analyze data on student test scores, graduation and attendance rates, and teacher competency levels. States must send the information to school districts, which then must provide it to parents. Schools are held accountable for making Adequate Yearly Progress (AYP). Serious sanctions are imposed upon schools not at the minimum "meets or exceeds" levels on the yearly state math assessment. And ironically, given the abovementioned *America 2000* goal, the first major aspect of *NCLB* is "every child proficient in math and reading by 2014." But what does "proficient" mean for a child with James' mathematical talents?

NCLB: LEAVING THE GIFTED BEHIND?

Improving the quality of the gifted has no rewards in No Child Left Behind. The philosophy of NCLB is quantitatively based towards the needs of the lowest achieving students. All (or at least the majority) of students need to "meet" a certain level of math proficiency. There are punishments for not meeting this goal yearly. However, there are no rewards for improving the "exceeds" students.

To avoid sanctions, schools are focusing their resources to raise lower-achieving children to be "proficient." To raise mathematics and reading test scores in lower-scoring students, other aspects of education are losing resources, including the opportunities for the gifted and talented. But serving the basic needs of the majority disenfranchises the needs of the gifted individual. What about the qualitative aspects of education? Tharman Shanmugaratnam, the Minister of Education of Singapore, commenting on the U.S. educational system as a failure, says the gifted U.S. students do not do well because of teaching methods that focus on bringing everyone along, thus the bright students are never pushed (*Newsweek*, Jan. 9, 2006). What about the qualitative aspects of education?

James' individual needs are not being served. James, and other gifted mathematics students, are not being allowed to be "self-proficient in math" at *their individual ability level*. It is this very small minority of extremely gifted students that the U.S. desperately needs to become the future scientific, medical, and technological leaders. Jan and Bob Davidson, in their book *Genius Denied: How to Stop Wasting our Brightest Young Minds*, refer to these children as a "natural resource that's being squandered" (Davidson & Davidson, 2004).

If James's story sounds familiar—it is. James is not the only student affected by the lack of an individualized gifted program. Thirteen years ago in 1994, it was Jeff, a freshman in high school—extremely gifted, academically underachieving, extremely frustrated, and dreadfully bored. Jeff's story was the impetus for the article, "Failing the Needs of the Gifted: The Argument for Academic Acceleration of Extremely Gifted Mathematics Students" (Ream & Zollman, 1994).

This article is a re-visitation of this 1994 article. What progress has been made? What setbacks have occurred? This paper on the needs of the mathematically gifted will: first, review the different types of gifted students; second, review approaches to fulfilling the needs of the gifted; third, discuss the current barriers with the needs of the extremely gifted; fourth, examine the opportunities available for the extremely gifted; and lastly, speculate on the future for gifted mathematics student in school.

DEFINING THE DIFFERENT TYPES OF "GIFTED" STUDENTS

Approximately 5% of the U.S. population is considered "gifted," with IQs of 115 and above. One out of a thousand people, 0.1% of the U.S. population, is considered highly gifted (IQ 145+), while extremely gifted people (IQ 160+) appear in the population at a rate of 1 in 10,000, 0.01% of the population (Davidson & Davidson, 2004).

A student that easily does well in mathematics in school is given the loose lay term, "gifted and talented" by most people (including teachers). But the literature points out that there are levels of giftedness (Greenes, 1981). One type of student regularly performs well in school. But these students do well because of hard work and perseverance. Greenes calls such students "*good exercise doers.*" Greenes argues that these students are misidentified as gifted.

The second type of students is "above" the good exercise doers group. They reason well, solve non-routine problems, learn new material quickly, retain the new material, and are able to apply and transfer their knowledge. They share a variety of characteristics in their mathematics methodology that other students do not (or cannot) do. These students can work independently for long spans of time. They reflect abstractly. Such students may be accurately labeled as "*highly gifted.*"

The third class of students consists of those who are highly precocious; who, with little or no formal instruction, who perform at the level of students several years older; who learn at a fast rate; and who deal well with sophisticated content and problems, "*extremely gifted.*" Sriraman (2005) adds another level above these, namely "*mathematically creative.*" Mathematically creative students produce (a) produce novel, insightful solutions to problems or (b) formulate new, imaginative questions to existing problems. This paper is concerned with students described by the second and third use of the label gifted, namely, "*highly gifted*" and "*extremely gifted.*"

Children in the last two types, *highly gifted* and *extremely gifted*, will be identified as gifted children in this paper. Gifted children excel mathematically; they are able to organize data, use a variety of approaches and strategies for solving problems, and are likely to find more than one solution to a problem (Sriraman, 2003). They frequently are reluctant to do busy work or practice skills already mastered. Instead, they enjoy challenges and complexity in mathematics and like to extend or make up their own problems. Long attention spans and the ability to work independently also characterize mathematically *highly gifted* and *extremely gifted* students. Their perfectionism and constructive criticism is often combined with a keen sense of humor. Stanley notes that although students who excel in mathematics have many intellectual traits in common, they are as different from each other in most personal and physical characteristics as are students of average ability of the same age (Stanley, 1977).

APPROACHES AND BARRIERS TO FULFULLING THE NEEDS OF THE HIGHLY GIFTED AND EXTREMELY GIFTED

Highly gifted and extremely gifted children in mathematics have special intellectual needs because of their unique interests and abilities. The two basic options for maintaining the high levels of interest and achievement of mathematically gifted children are (a) academic acceleration and (b) academic enrichment. Academic acceleration involves moving gifted children through the standard curriculum more quickly than students of average ability. Academic enrichment involves gifted students studying the standard curriculum at the same pace as average students, but at a broader and deeper level.

The National Council of Teachers of Mathematics (NCTM) addresses the use of these options. NCTM recommends, "all mathematically talented and gifted students should be enrolled in a program that provides a broad and enriched view of mathematics in a context of higher expectation" (House, 1987, p. 100). NCTM endorses the Schoolwide Enrichment Model (SEM) (Renzulli & Reis, 2003) and the use of the Renzulli Enrichment Triad Model (House, 1987). NCTM contends that, in almost all cases, gifted students benefit more from enrichment than acceleration (Sheffield, 1994). However, for a limited number of "extremely talented and productive [students]...whose interests, attitudes, and participation clearly reflect the ability to persevere and excel throughout the entire program," NCTM supports accelerated programs enhanced by enrichment (House, 1987; NCTM, 2000).

Federal funding for gifted education is small. The U.S. Department of Education reported 2 cents of every $100 (0.02%) spent on education goes toward gifted programs. The government program is called the Jacob K. Javits Gifted and Talented Students Education Program. The purpose of this program is to carry out a coordinated program of scientifically based research, demonstration projects, innovative strategies, and similar activities designed to build and enhance the ability of elementary and secondary schools to meet the special education needs of gifted and talented students. On the current U.S. Department of Education's Jacob K. Javits website is the following: "Due to FY 2006 budget constraints, a new discretionary grant competition will not be held this year for the Jacob K. Javits Gifted and Talented Students Education Program. Future grant competitions are contingent upon available funding" (Javits, 2007).

The most conventional and least costly option for schools, academic acceleration, usually is the only option offered to the gifted, if any option is offered at all. However, academic acceleration alone does not meet all the needs of the highly and extremely gifted. Academic acceleration is doing "something" for the *intellectual needs* of the child; so most parents usually accept this. But academic acceleration does not meet other needs of the child, the emotional needs, the esteem needs, and the self-actualization needs (Maslow, 1954). According to Abraham Maslow, all humans have five levels of needs, as shown below:

- Level 5: *Self-Actualization Desires*—self-fulfillment, seeking personal development. [B-need]
- Level 4: *Esteem Needs*—self-esteem, achievement, mastery, independence, status, dominance, prestige, managerial responsibility, influence. [D-need]
- Level 3: *Emotional Needs*—sense of belonging and love in work group, family, relationships. [D-need]
- Level 2: *Safety Needs*—security, order, law, limits, stability. [D-need]
- Level 1: *Survival Needs*—food, drink, shelter, warmth, sleep. [D-need]

These human needs are in a hierarchy of levels. Needs must be satisfied in the given order, from level 1 up to level 5. Emotional and esteem levels (3 and 4) Maslow calls *deficit needs*, or D-needs (along with two lower levels, namely, survival needs (1) and safety needs (2)). If you don't have enough of something—i.e. you have a deficit—you feel the need. But if you get all you need, you feel nothing. In other words, the needs cease to be motivating. The last level, self-actualization, is a different. Maslow has used a variety of terms to refer to this level: He has called it growth motivation (in contrast to deficit motivation), *being needs* (or B-needs, in contrast to D-needs), and self-actualization (Maslow, 1954).

Self-actualization needs are needs that do not involve balance or homeostasis. Once engaged, they continue to be felt. In fact, they are likely to become stronger as they are "feed." They involve the continuous desire to fulfill potentials, to "be all that you can be." They are a matter of becoming the most complete, the fullest, "you"—hence the term, self-actualization. Thus, in keeping with Maslow's hierarchy of human needs theory up to this point, if you want to be truly self-actualizing, you need to have your lower needs taken care of, at least to a considerable extent (Maslow, 1954).

Again, schools opt to offer only academic acceleration, either by "promoting early" elementary students, scheduling middle and high school students into higher levels of mathematics or special honors/advanced placement courses. Research in acceleration show mainly positive advantages for the gifted (Ablard, Mills, & Duvall, 1994; Brody, Assouline, & Stanley, 1990,

Kolitch & Brody, 1992). Negative effects also are seen, chiefly feelings of isolation and un-comfort. These did not have a great impact according to Ablard, et al. (1994), "because the opportunity to be intellectually challenged often outweigh any social disadvantages."

Academic acceleration, for a time, can fulfill these students' needs of self-esteem. But even these students (with emotional needs fulfilled by the home, and esteem needs fulfilled by the school through academic acceleration) are still at an extreme risk of not reaching their level of self-actualization and personal development. These students require specially designed courses, with specially trained teachers, to challenge them academically for their "being needs" (B-needs); and to support them emotional to satisfy their "deficit needs" (D-needs). Otherwise, the students risk becoming the "gifted underachievers" with hostility towards school, poor study habits, and low levels of self-discipline.

OPPORTUNITIES AVAILABLE FOR THE HIGHLY AND EXTREMELY GIFTED

There are opportunities for the highly and extremely gifted that go beyond mere academic acceleration. But these opportunities are not easily found in the public school and are quite limited. Even "magnet schools" (specially academically-designated schools) are more for the "good exercise doers" than the highly and extremely gifted. Daria Hall, Assistant Director of the Education Trust, says their review of transcripts shows evidence of "course inflation"—offering high-level courses that have the "right names" but a "dumbed-down" curriculum (*LA Times*, Feb. 23, 2007).

Some lucky students are in a state, Illinois for example, that has *one* selective mathematics and science academy, a boarding high school for the gifted and talented. These academies are limited to a very small number of high school students that are bright enough, usually affluent enough, and lucky enough to get accepted.

In the U.S., a select number (approximately 20) of states and universities have camps, institutes and summer programs for the gifted student. Most of these are expensive, as are most types of summer camps. Finding out about these specialized camps opportunities is not widely known, even to high school teachers and counselors. A greater number of universities allow gifted high school students to come on campus and take university classes. Again, this is a fairly expensive option, not widely available to lower and middle class families. Four major universities (Duke, Johns Hopkins, Northwestern, and Denver) have gifted and talented search programs. And Johns Hopkins and Stanford Universities have specific on-line courses for gifted children.

Outside of this very small number of high school students allowed in these limited options, parents are "on their own" in finding resources to fulfill the needs of the gifted student. The Internet has become the major source of information and resources for the parents of the gifted. The National Research Center on the Gifted and Talented, at the University of Connecticut, and the American Association of Gifted Children, at Duke University, maintain active Internet web sites of pertinent information for parents. Specialized sites, run mostly by volunteers, such as Genius Denied, Davidson Institute for Talent and Development, Resources for Gifted Families, Highly Gifted, and Twice Exceptional Hoagies' Gifted Education, give a voice to parents of gifted children (Rotigel & Fello, 2004).

There also are Internet sites designed specifically for use by students. These are not solely for the gifted student. Such sites as Math Forum, Go-Math, Mathlab, and How Stuff Works, provide enrichment activities and problems for any interested student that has the ability to work individually on tasks without supervision (Rotigel & Fello, 2004).

THE FUTURE FOR HIGHLY AND EXTREMELY GIFTED MATHEMATICS STUDENT IN SCHOOL

Most articles on the gifted begin with a personal story of a child. Why? The gifted are not the majority of students, with the highly and extremely gifted being 0.1% of the student population. The story of the needs of the gifted is an individual story of the "waste" of our country's brightest and best. The mathematically highly and extremely gifted are but a few in number. "Losing" one is significant. (Indeed, China and India's top 12.5% of students nearly equals the number of *total* U.S. students.)

In the U.S., physically and mentally challenged students have federal regulations that require schools to provide for their individual needs. Every challenged student, by law, has an Individual Educational Plan (IEP) specific designed for that student's emotional needs, social requirements, personal desires, and individual abilities. The gifted have no protection for their special needs. Further, it is not feasible to believe that the gifted student will demand specialized treatment. In fact, most students try not to be designated differently than their peers. (This is a level 3 need, their need to belong.)

Opportunities for the gifted student do exist, but they must be sought out and demanded by the guardian for the child. Within the traditional school, the gifted are, at best, only academically accelerated. Outside of the school, limited opportunities for the gifted are available—if the child's guardian is knowledgeable, affluent, and lucky.

This article began with the question, "Are gifted students surviving or thriving?" Revisiting the situation in mathematics the answer is too often that it depends of the affluence, awareness, and the tenacity of the parents. The public school has concerns other than the individual gifted student's needs. Our country, our world, needs all students, and particularly the highly and extremely gifted, to reach their self-actualization. These students are rare and valuable resource.

REFERENCES

Ablard, K. E., Mills, C. J., & Duvall, R. (1994). *Acceleration of CTY math and science students.* Baltimore, MD: Johns Hopkins University, Center for Talented Youth.

America 2000: An Education Strategy. (1991). Washington, D.C.: U.S. Department of Education.

Brody, L. E., Assouline, S. G., & Standley, J. C. (1990). Five years of early entrants: Predicting successful achievement in college. *Gifted Child Quarterly, 34,* 138–142.

Davidson, J., & Davidson, R. (2004). *Genius denied: How to stop wasting our brightest young minds.* New York: Simon & Schuster.

Greenes, C. (1981). Identifying the gifted student in mathematics. *Arithmetic Teacher, 28*(6), 14–17.

House, P. (Ed.). (1987). *Providing opportunities for the mathematically gifted K–12.* Reston, VA.: National Council of Teachers of Mathematics.

Jacob K. Javits Gifted and Talented Students Education Program. (2007). Available: http://www.ed.gov/programs/javits/index.html.

Kolitch, E. R., & Brody, L. E. (1992). Mathematics acceleration of highly talented students: An evaluation. *Gifted Child Quarterly, 36*(2), 78–85.

Landsberg, M. (2007, February 22). Study says students are learning less. *Los Angeles Times.*

Maslow, A. H. (1954). *Motivation and Personality.* New, York: Harper & Row.

National Center for Education Statistics. (1999). *Third International Math and Science Survey.* Washington DC: U.S. Department of Education.

National Center for Education Statistics. (2003). *Trends in International Math and Science Study.* Washington DC: U.S. Department of Education.

National Council of Teachers of Mathematics (2000). Principles and standards for school mathematics. Reston, VA: Author.

Ream, S. K., & Zollman, A. (1994). Moving on: The case for academic acceleration of extremely gifted mathematics students. *Focus on Learning Problems in Mathematics, 16*(4), 32–43.

Renzulli, J. S. & Reis, S. M. (2003). The Schoolwide enrichment model: Developing creative and productive giftedness. In Colangelo, N. and Davis, G. A. (Eds.). *Handbook of Gifted Education* (184–203). Boston: Allyn and Bacon.

Rotigel, J. V., & Fello, S. (2004). Mathematically gifted students: How can we meet their needs? *Gifted Child Today, 28*(4). 46–51.

Sheffield, L. (1994). *The development of gifted and talented mathematics students and the National Council of Teachers of Mathematics Standards.* Storrs: National Research Center on the Gifted and Talented, University of Connecticut.

Sriraman, B. (2003). Mathematical giftedness, problem solving, and the ability to formulate generalizations. *The Journal of Secondary Gifted Education. 14*(1), 151–165.

Sriraman, B. (2005). Are giftedness and creativity synonyms in mathematics? *The Journal of Secondary Gifted Education. 17*(1), 20–36.

Stanley, J. (1977). Rationale for the Study of Mathematically Precocious Youth (SMPY) during its first five years of promoting educational acceleration. In J. Stanley, W. C. George, & C. H. Solano (Eds.), *The gifted and the creative: A fifty year perspective.* (pp. 75–112). Baltimore: Johns Hopkins University Press.

U.S. Department of Education. (2001). *No child left behind act of 2001* (Public Law 107-110). Washington, DC: author.

Zakaria, F. (2006, January 9). We all have a lot to learn. *Newsweek.*

ACKNOWLEDGMENT

Reprint of Zollman, A. (2007). Revisiting the needs of the gifted mathematics students: Are students surviving or thriving? In B. Sriraman (Guest Editor), *Mediterranean Journal for Research in Mathematics Education, 6*(1 & 2),139–148. Reprinted with permission from The Cyprus Mathematical Society. ©2007 Alan Zollman.

CHAPTER 14

PLAYING WITH POWERS

Bharath Sriraman
The University of Montana

Pawel Strzelecki
Institute of Mathematics, Warsaw University, Poland

ABSTRACT

This paper explores the wide range of pure mathematics that becomes accessible through the use of problems involving powers. In particular we stress the need to balance an applied and context based pedagogical and curricular approach to mathematics with the powerful pure mathematics beneath the simplicity of easily stated and understandable questions in pure mathematics. In doing so pupils realize the limitations of computational tools as well as gain an appreciation for the aesthetic beauty and power of mathematics in addition to its far-reaching applicability in the real world.

INTRODUCTION

In this paper we explore the curricular and pedagogical implications of the general area of "powers." By "powers" we mean the mathematics arising from the study of exponential functions. In the United States, the Algebra

Creativity, Giftedness, and Talent Development in Mathematics, pages 287–297
Copyright © 2008 by Information Age Publishing
All rights of reproduction in any form reserved.

Standard recommends that students in grades 9–12 be introduced to the exponential function and classes of functions in general including polynomial, logarithmic and periodic functions (NCTM, 2000). Traditionally 9th and 10th graders (13–16 year olds) in the U.S are introduced to exponential functions by first extending the laws of exponents from integer powers to real powers. This is usually followed by graphing exponential functions such as 2^x, 3^x, etc. in order to convey the fact that for functions of the form a^x ($a > 1$), the domain is the set of real numbers whereas the range is $(0, \infty)$. The graph also reveals properties such as the increasing behavior of the function and the X-axis being the horizontal asymptote for this function as $x \to -\infty$. Students are then introduced to the case where $0 < a < 1$ to study properties that we are all familiar with. A capstone topic in this traditional treatment of exponential functions is introducing students to the irrational number 'e' by studying the behavior of $(1 + 1/n)^n$ as $n \to \infty$. Finally students are introduced to logarithms and the analogous laws of logarithms are derived from the laws of exponents.

In the mathematical experiences of the second author in Poland, as a high school student, exponential functions appeared a tiny bit later in the curriculum in comparison to the U.S., at the beginning of the 3rd year of secondary school (for students at the age of 16 to 17). The sequence of events was similar to the one described in the previous paragraph. First, students learned how to define powers with rational exponents so that all laws that hold for natural exponents were still satisfied. Then, looking at graphs, they were prompted to note that exponential functions map arithmetic progressions to geometric ones (and to use this property for graphing exponentials on graph paper) etc. Finally, powers with arbitrary real exponents were defined in a more or less precise way, using monotonicity. It is hard to say precisely how many students were able to digest this section of the textbook and to retain the comprehension. Not too many, we are afraid. In our opinion, the aim of the textbook was clearly, to purify and to "bourbakize"... alas, with no real success. The Bourbaki[1] were a group of mostly French mathematicians, who began meeting in the 1930s and aimed to write a thorough (formalized) and unified account of all mathematics. The "bourbakized" definition of $2^{\sqrt{2}}$ as the supremum of a suitable set of rational powers of 2 is not really entertaining, nor enlightening! Even mathematically oriented 16-year-olds have lots of more interesting things to do! Pure mathematics should be, among other things, a source of fun—we shall come back to this later.

In the traditional curriculum students also encounter word problems associated with exponential functions. Functions obtained by running an exponential regression on data arising from the natural, social and financial sciences are given *a priori* to the students in the context of solving these word problems. Numerous examples of such functions abound. For instance $p = 760e^{-0.145h}$ relates the pressure p on a plane (in millimeters of mercury) to height h (in kilometers) above sea level. The spread of information via mass media (TV and magazines) is modeled by $d = P(1 - e^{-kt})$ where P stands for a fixed population and t denotes time (Coleman, 1964). Classical examples are of course found in physics. For instance Newton's law of cooling is given by $f(t) = T + (f_0 - T)e^{kt}$, where $k < 0$, f_0 is the initial temperature of the object that is heated, T is the temperature of the surrounding medium. Other classical examples are radioactive decay, growth of bacteria, etc.

This approach has been reversed by reform-based curricula in the U.S. that emphasize modeling activities as a means to generate data. For instance bouncing balls with a known co-efficient of elasticity can be used. Students are asked to drop a ball (basketball, tennis ball, etc.) from a starting height and asked to keep track of the bounce height as the bounces progress until the ball is flat on the ground. Such data is made sense of by realizing that the plot of bounce height over bounce numbers shows an exponentially decaying pattern and therefore results in the invoking of an exponential regression. For instance the data for a bouncing basketball is modeled fairly accurately by $f(x) = A(0.5)^x$ where A is the initial height and x is the bounce number. These are specific examples of the wider range of model eliciting activities (Lesh & Doerr, 2003) that expose students to applied mathematics in a real world context very early in their schooling experiences. This is in stark contrast to the "good old" days when one had to take a course in differential equations to encounter such modeling problems in a "recipe driven" didactic environment.

Exposure to the applied aspects of mathematics conveys to students only one aspect of the spectrum of mathematics. We claim that pure math activities can also be accomplished by problems involving powers, which—despite their surprisingly elementary statements—can lead students to deep insights into the behavior of numbers and the exciting possibilities in pure mathematics. Our goal in this paper is to demonstrate these alternative possibilities in pure mathematics by playing with powers, thus broadening the spectrum of the students.

POWER PROBLEMS AND REMAINDERS: LEAD-INS TO NUMBER THEORY BASICS

Students encounter basic number theory concepts such as prime and composite numbers, the division algorithm, divisibility tests, notions of least common multiple and greatest common divisor and the fundamental theorem of arithmetic in the middle school grades [10–13-year-olds]. Unfortunately there is little or no follow up to these topics in the higher grades.

One of the curricular flaws of viewing Calculus as the pinnacle of the students' secondary schooling experience is working pre-dominantly over the set of real numbers with functions of continuous variables. The traditional sequence of Algebra-Geometry-Trigonometry and Analytic Geometry offers little opportunity to further develop elementary number theory notions that students have previously encountered. Problems involving powers can be very useful to remedy this unfortunate situation. Most of the extant books that include a treatment of problems involving powers and number theory are classified as "competition" books, thereby creating an "elitist" aura around such problems. We think otherwise and invite teachers to make appropriate use of "power" tasks to convey to students the power of pure mathematics and the limitations of computing tools. An illustration of such a problem follows.

One can pose the question: What is the remainder when we divide 6^{131} by 215? Most calculators would not be able to handle 6^{131}, which immediately necessitates us to examine the problem with different conceptual tools. We can reflect on what remainders mean? We can go back to notions from the early grades and start basic computations such as:

> What is the remainder when we divide 5 by 3? It's 2. That was easy.
> What is the remainder when we divide $5 \times 8 = 40$ by 3? It's 1. Still easy!
> Now what is the remainder when we divide $5 \times 8 \times 7 = 280$ by 3? We could laboriously perform the division by 3 and find that the remainder is 1.

Is there a lesson to be learned through this empirical work? Yes! The important mathematical phenomenon to communicate is that if we calculate the remainders of 5, 8, and 7 (which are 2, 2, and 1 respectively) and multiply these individual remainders $2 \times 2 \times 1 = 4$ and divide this by 3, we get the same remainder when dividing 280 by 3. This phenomenon is in general called the Remainder theorem, which can simply be stated as: the remainder of a product of numbers divided by a given number is equal to the remainder of the product of the remainders of the numbers (constituting the product) divided by this given number.

Now getting back to the original problem. We can employ the laws of exponents that our students so faithfully memorize and write 6^{131} as a product of numbers that leave a "nice" remainder when divided by 215. The best candidate is 6^3 since $6^3 = 216$ and 216 divided by 215 leaves a remainder of 1, and we can easily multiply 1s.

So $6^{131} = 6^3 \times 6^3 \times 6^3 \times 6^3 \ldots$ (43 times) $\times 6^2$ divided by 215 leaves a product of remainders $1 \times 1 \times 1 \times 1 \ldots$ (43 times) $\times 36$, which is 36 and 36 divided by 215 leaves a remainder of 36. Done! This problem not only allowed us to make use of laws of exponents but also led to some insights into the theory of numbers.

Another nice problem involving powers is calculating the last digit of a given power, typically a huge number, such as 777^{777} or 2004^{2004}. Once again this allows us to make a nice foray into the basic behavior of numbers. If we consider a simpler problem such as 1^{12345} we observe that the last digit is obviously 1 since we are simply multiplying a product of 1s. What if we had 11^{12345}, we can empirically verify that 11^1, 11^2, 11^3, 11^4... always end in 1. Students can be led to observe that the last digit is 1, and in general the last digit of any huge number such as 777^{777} is the product of the last digit multiplied by itself the given number of times. So the last digit of 777^{777} is the same as the last digit of 7^{777}. Now if we write out the powers of 7, we find a periodicity phenomenon in the last digits related to its powers. This is observed by listing out the sequence of the powers of 7: 7^1, 7^2, 7^3, 7^4, 7^5, 7^6, 7^7, 7^8, 7^9... which gives last digits 7, 9, 3, 1, 7, 9, 3, 1, 7.... So we can once again employ laws of exponents to rewrite 7^{777} as:

$$7^{777} = (7^4) \times (7^4) \times (7^4) \times (7^4) \times \ldots (194 \text{ times}) \times 7^1$$

so the last digit is 7.

Since we seem to be able to determine last digits of ridiculously large numbers, why not look at the problem of determining first digits.

STARTING DIGIT PROBLEMS: "EXCUSES" INTO COMBINATORICS AND ANALYSIS

Let's start with an existence problem. Is there an integral power of 2 that begins with 1999... in its decimal expansion? In other words the question is asking us to prove the existence of an integer 'n' such that $2^n = 1999$... without explicitly asking exactly what this power is. We can clearly assume that $n > 0$. When we raise 2 to any integer power, there are 9 obvious choices for the first digit since we naturally exclude zero, and then there are 10 choices for each digit after that. So there are $9 \cdot 10 \cdot 10 \cdot 10 = 9000$ possible ways of listing the first four digits. Since $n > 0$, we have no restrictions on the number of values we can generate, and we can easily generate more than 9000 values for 2^n. By the pigeonhole principle,[2] some powers of 2 have to begin with the same four-digit string but we are still not sure that one of those starting strings really equals 1999. This is a subtle question with an intriguing relation to irrationality and we postpone it for a second to mention something simpler.

A nice problem that is easily solved via the pigeonhole principle and makes use of divisibility properties is to prove that there exists some power of 3 that ends in 001. This can easily be shown as follows. Suppose 3^m and 3^n (where $m > n > 1$) upon division by 1000 have the same remainder. (The existence of such m and n needs to be established by applying the pigeonhole principle and we will leave it up to the readers.) Then $3^m - 3^n = 3^n(3^{m-n} - 1)$ is divisible by 1000. Now 3^n and 1000 clearly have no common factors, which means 1000 has to divide the factor $(3^{m-n} - 1)$. This implies 3^{m-n} ends with 001.

Now let us come back to powers of 2, and to the question whether one of them begins, in decimal notation, with 1999. Maybe this string is too strange to appear at the beginning of the decimal notation of some power of 2? Let us try something simpler first. Consider a sequence $a(n)$ consisting of the first digits of the consecutive powers of 2:

1, 2, 4, 8, 1, 3, 6, 1, 2, 5, 1, 2, 4, 8, ... Will 7 ever appear in this sequence?

This problem in different versions is known in mathematical literature. The first mention is found in the famous book *Ordinary Differential Equations* (Arnold, 1978). It is usually accompanied by auxiliary facts or suggestions intended to make a solution more accessible. Nevertheless, we have not encountered any detailed solutions to this problem. We will remedy this unfortunate situation and in the process illustrate the rich mathematics that comes out of it. To start with—a desperate solution using a piece of paper and a pencil or another powerful computing tool will easily verify that

$$2^{46} = 70,368,744,177,664$$

Going further with this experiment we can see that 7 is the first digit of 56th, 66th, 76th, 86th and the 96th power of 2 (but the first digit of 106th power of 2 is 8, not 7). This empirical method, however, is clearly far from being mathematically elegant. We need a better solution, which would allow us to draw other conclusions. First of all, we must try to realize the meaning of the statement that 7 is the first digit of a number 2^n. The answer is simple: 7 is the first digit of 2^n if and only if for some natural k we have

$$7 \cdot 10^k < 2^n < 8 \cdot 10^k$$

We can get a simpler description of this condition if we take the decimal logarithms of both sides, something our students would be quite familiar with. This yields $k + \log(7) < n \log(2) < k + \log(8)$. Since decimal logarithms of 7 and 8 lie between 0 and 1, we conclude that k is the integer part of the number $n \log(2)$, which leads to the following inequalities:

$$\log 7 < n \log 2 - [n \log 2] < \log 8$$

And now it suffices to bring together some known facts, which we invite the readers to verify.

Lemma 1: *The number log(2) is irrational.*

Lemma 2: *If a number x is irrational and c(n): = nx – [nx], then for any a and b belonging to [0, 1] infinitely many members of the sequence c(n) lie in the interval (a, b).*

We hope that readers will encourage their students to prove lemma 1, which is very easily provable via contradiction. Assuming this done, let us have a look at a proof of the second lemma and then examine their consequences.

Proof of Lemma 2. Observe first that all the members of the sequence $c(n)$ are different. Indeed, if $c(k) = c(m)$ for $k > m$, then $(k - m)x = [kx] - [mx]$. This is a contradiction, since the product of a non-zero integer $(k - m)$ and an irrational number x cannot be an integer.

Take now a positive integer n such that $1/n < b - a$. The numbers $c(1)$, $c(2), \ldots, c(n + 1)$ being all different and belonging to the interval $[0, 1]$, we infer by the pigeon-hole principle that for some i and s such that i and $(i + s)$ both lie between 1 and $(n + 1)$ the following inequality holds:

$$0 < \varepsilon = \left| c(i) - c(i + s) \right| \leq 1/n < b - a \tag{14.1}$$

Now wrap the real axis into a circumference **T** of perimeter 1 with a distinguished point 0. For two numbers a and b in $[0, 1]$ we denote by (a, b) the arc of **T** which corresponds to the interval (a, b) on the real axis. Let $f: \mathbf{T} \to \mathbf{T}$ be an anti-clockwise revolution by an angle of $2\pi x$ radians. Instead of watching the numbers $c(n)$ in the interval $[0, 1]$ we shall look at the images of the distinguished point 0 under iterations of f on **T**. After a moment of reflection we note that the length of an arc $(0, b(n))$, where $b(n) = f^n(0) = f \bullet f \bullet \ldots \bullet f(0)$ [$\bullet$ = composition of mappings] is equal to $c(n)$ Hence, due to (1) we know that the length of an arc between the points $b(i)$ and $b(i + s)$ is smaller than $b - a$. This means that the sth iterate of f is a revolution by an angle of $2\pi\varepsilon$ radians; the direction of this revolution is of no importance to us. This obviously implies that infinitely many of the points $b(s), b(2s), b(3s), \ldots$ belong to the arc (a, b). Indeed, if we start from the fixed point 0 and walk for an infinitely long time along the circumference **T** in one and the same direction making steps of length ε, then infinitely many times we shall step into the arc (a, b), since its length $b - a$ is greater than the length of our step. This completes the proof.

Now, by applying Lemma 2 to $x = \log(2)$, $a = \log(7)$, $b = \log(8)$ we get that 7 is the first digit of infinitely many powers of 2. If we apply again

Lemma 2 to numbers $x = \log(2)$, $a = \log(77)-1$, $b = \log(78)-1$, then because of the equalities $1 = [\log 77] = [\log 78]$, we conclude that the figure seven can even appear twice in the first two places of the decimal notation of a power of 2. Using an analogous argument we readily discover that any finite sequence of digits can appear at the beginning of a decimal notation of a power of 2, like 1234 or 567890, or (finally) 1999. If you really do not believe this last statement, we invite you to compute (say) 2^{9030} or 2^{11166}. At the end of this article we exhibit some important historical dates compared with corresponding powers of 2—to satisfy the skeptics. We can further deduce the following corollary:

> **Corollary:** *If an integer $p > 1$ is not an integer power of 10, then any sequence of digits can appear at the beginning of the decimal notation of n-th power of p for some n.*

So the question again is why does 7 not appear among the first members of the sequence we introduced at the very beginning? Why does this deceitful sequence pretend to be periodical? The reason is simple. The number $\log(2) = 0.3010299956\ldots$ can be very well approximated by the rational number 0.3, and for all rational x the sequence $c(n) = nx - [nx]$ is periodical. In other words: $2^{10} = 1024$, which is quite close to 1000. And multiplication by 1000 just adds zeroes at the end; front digits stay unchanged. This is why after seeing the first few members of the sequence $a(n)$ we come to an erroneous conclusion that the sequence has period 10 and 7 is not a member of it, while 8 appears quite often. To see the first 7 one has to look at the first 6 in $a(n)$ (i.e., the first digit of $64 = 2^6$), and then wait until the cumulative effects of small perturbation $24 = 2^{10}-1000$ do their job.

In 1910 Sierpinski, Weyl, and Bohl proved independently of one another that for every irrational x the sequence $c(n) = nx - [nx]$ is equidistributed over the interval $[0,1]$ (Arnold & Avez, 1968). More precisely, if we take arbitrary a and b (with $a < b$) from $[0,1]$, and let $k(n, a, b)$ denote the number of elements of the set $\{c(i): 1 \le i \le n, c(i) \in (a, b)\}$ then we get:

$$\lim_{n \to \infty} k(n, a, b)/n = b - a$$

Speaking in more illustrative terms, this theorem states that if we walk along a circle of circumference 1 unit, taking steps of irrational length, then we step on each hole with a frequency which is directly proportional to the size of the hole! Let's translate this fact into our language of powers of 2. Let $a(7, n)$ and $a(8, n)$ be the number of sevens and eights, correspondingly, among the first *n* members of the sequence $a(n)$. By the last formula, we have

$$\lim_{n \to \infty} a(7, n)/n = \log 8 - \log 7$$

$$\lim_{n \to \infty} a(8, n)/n = \log 9 - \log 8, \text{ so consequently}$$

$$\lim_{n \to \infty} a(7, n)/a(8, n) = (\log 8 - \log 7)/(\log 9 - \log 8) = 1.1337\ldots > 1$$

This means that by looking at sufficiently long initial fragments of the sequence $a(n)$ we will see slightly more sevens than eights. This result of Bohl, Sierpinski, and Weyl, which we previously mentioned and our little fact on 7s and 8s are in fact simple consequences of a very general and deep theorem in ergodic theory due to G. D. Birkhoff (Cornfeld, Fomin, & Sinai, 1982), which the interested reader is urged to pursue.[3] We conclude this section with a little problem for our readers.

> **Problem:** *For what n does the number 2^n have four consecutive sevens at the beginning? What about five sevens? How can we estimate from above the least n such that the decimal notation of 2^n begins with 2004 consecutive sevens?*

FUNCTIONAL PROBLEMS: GETTING DEEPER

If we play with the laws of powers (exponents), then we observe that $a^{x+y} = a^x \cdot a^y$. Let us assume that $a > 0$. The question can be posed generally as: Suppose $f: \mathrm{R} \to \mathrm{R}$, what are all the other solutions to $f(x + y) = f(x) \cdot f(y)$? Another good problem arises from observing that $\log(xy) = \log x + \log y$. So what are all the other solutions to $f(x \cdot y) = f(x) + f(y)$? A classic related problem is of course to find all functions that satisfy the Cauchy functional equation: $f(x + y) = f(x) + f(y)$. In answering these problems one gets into a deep investigation of functional equations, something that teachers can use as an extended project for the motivated and bright students.

SOME UNUSUAL POWERS OF 2 AND CONCLUSIONS

In order to satisfy the skeptics of the unusual property of powers of 2 we list some powers of 2 that connect with a (biased) sample of important historical dates.

The baptism of Poland	966	$2^{568} = 9.66\ldots \times 10^{170}$
The battle of Hastings	1066	$2^{5561} = 1.066\ldots \times 10^{1674}$
Columbus discovers America	1492	$2^{3761} = 1.492\ldots \times 10^{1132}$
The founding of Harvard University	1636	$2^{9528} = 1.636\ldots \times 10^{2868}$
Cromwell's death	1658	$2^{3223} = 1.658\ldots \times 10^{970}$
The founding of Royal Society	1660	$2^{4874} = 1.660\ldots \times 10^{1467}$
New Amsterdam changes name to New York	1664	$2^{6040} = 1.664\ldots \times 10^{1818}$
First edition of Newton's "Principia"	1687	$2^{6143} = 1.687\ldots \times 10^{1849}$
Walpole becomes Britain's first Prime Minister	1721	$2^{10229} = 1.721\ldots \times 10^{3079}$
French Revolution	1789	$2^{9857} = 1.789\ldots \times 10^{2967}$
Waterloo	1815	$2^{931} = 1.815\ldots \times 10^{280}$
Beginning of World War II	1939	$2^{5522} = 1.939\ldots \times 10^{1662}$
End of World War II	1945	$2^{1931} = 1.945\ldots \times 10^{581}$

We hope to have conveyed to the readers the richness of pure mathematics present in problems involving powers. Playing with problems involving powers to make deep connections with topics in Number Theory, Combinatorics and Analysis complements the Applied mathematics and Statistics that students learn through the modeling approach that is presently gathering momentum. The first author used first and last digit problems similar to the ones mentioned in this paper with 14-year-old pupils enrolled in an Algebra course. The pedagogical goal was to mediate "pure math" problem solving experiences and resulted in considerable student interest in the mysteries of the integers. Among other things, students realized the limitations of computing tools and understood the need to create/invent conceptual tools to tackle the problem. Other problems involving a particular phenomenon among the positive integers and resulting in the discovery of the pigeonhole principle also met with great success in the classroom (Sriraman, 2004a; 2004b).

In conclusion, our hope is that we never forget the aesthetic beauty inherent in pure math activities and convey to our students that such activities have sustained the imagination of mathematicians and contributed to its growth from the very onset of its history. We feel that the image of a pure mathematician lying under a tree (apparently doing "nothing" to the untrained eye!) complements and balances the image of the diligent applied mathematician and scientist engrossed in making sense of the hubris and chaos of the real world. After all, what would the second person do if the first one really did nothing?

REFERENCES

Arnold, V. I. (1978) *Ordinary differential equations* (translated from Russian by R. A. Silverman). Boston, MA: MIT Press.

Arnold, V. I., & Avez, A. (1968) *Ergodic problems in classical mechanics,* New York: Benjamin.

Coleman, J. (1964). *Introduction to mathematical sociology.* New York: The Free Press.

Cornfeld, I., Fomin, S., & Sinai, Ya. G. (1982). *Ergodic theory.* New York: Springer-Verlag.

Lesh, R., & Doerr, H. (2003). Foundations of a models and modeling perspective on mathematics teaching, learning and problem solving. In R. Lesh & H. Doerr (Eds.), *Beyond constructivism* (pp. 3–34). Mahwah, NJ: Erlbaum.

National Council of Teachers of Mathematics. (2000). *Principles and standards for school mathematics.* Reston, VA: Author.

Sriraman, B. (2004a). Discovering a mathematical principle: The case of Matt. *Mathematics in School, 33*(2), 25–31.

Sriraman, B. (2004b). Reflective abstraction, uniframes and the formulation of generalizations. *The Journal of Mathematical Behavior, 23*(2), 205–222.

ACKNOWLEDGMENT

Reprint of Sriraman, B., & Strzelecki, P. (2004). Playing with powers. *The International Journal for Technology in Mathematics Education, 11*(1), 29–34. Reprinted with permission from Research Information Ltd & Bharath Sriraman ©2004 Bharath Sriraman. Article is also available without cost on the website of *International Journal for Technology in Mathematics Education,* Volume 11, No 1.

NOTES

1. The Bourbaki essentially aimed to write a body of work based on a rigorous and formal foundation, which could be used by mathematicians in the future. For more information see Bourbaki, N. (1970). *Théorie des Ensembles dela collection elements de Mathématique,* Hermann, Paris. The internet savvy can also refer to the Bourbaki website located at http://www.bourbaki.ens.fr/

2. The pigeonhole (or Dirichlet) principle states that if we have "m" pigeons and "n" pigeonholes, where $m > n$, then some pigeonhole contains more than one pigeon. This seemingly obvious principle has wide ranging applicability in mathematics.

3. An online reference with the specifics of this theorem is found in *MathWorld—A Wolfram Web Resource* at http://mathworld.wolfram.com/BirkhoffsErgodicTheorem.html

Printed in the United States
130573LV00002B/124/P